Materials Horizons: From Nature to Nanomaterials

Series Editor

Vijay Kumar Thakur, School of Aerospace, Transport and Manufacturing, Cranfield University, Cranfield, UK

Materials are an indispensable part of human civilization since the inception of life on earth. With the passage of time, innumerable new materials have been explored as well as developed and the search for new innovative materials continues briskly. Keeping in mind the immense perspectives of various classes of materials, this series aims at providing a comprehensive collection of works across the breadth of materials research at cutting-edge interface of materials science with physics, chemistry, biology and engineering.

This series covers a galaxy of materials ranging from natural materials to nanomaterials. Some of the topics include but not limited to: biological materials, biomimetic materials, ceramics, composites, coatings, functional materials, glasses, inorganic materials, inorganic-organic hybrids, metals, membranes, magnetic materials, manufacturing of materials, nanomaterials, organic materials and pigments to name a few. The series provides most timely and comprehensive information on advanced synthesis, processing, characterization, manufacturing and applications in a broad range of interdisciplinary fields in science, engineering and technology.

This series accepts both authored and edited works, including textbooks, monographs, reference works, and professional books. The books in this series will provide a deep insight into the state-of-art of Materials Horizons and serve students, academic, government and industrial scientists involved in all aspects of materials research.

Review Process

The proposal for each volume is reviewed by the following:

1. Responsible (in-house) editor
2. One external subject expert
3. One of the editorial board members.

The chapters in each volume are individually reviewed single blind by expert reviewers and the volume editor.

Arnab Chanda · Sarabjeet Singh Sidhu ·
Gurpreet Singh
Editors

Materials for Biomedical Simulation

Design, Development and Characterization

Springer

Editors
Arnab Chanda
Centre for Biomedical Engineering
Indian Institute of Technology Delhi
New Delhi, India

Sarabjeet Singh Sidhu
Department of Mechanical Engineering
Sardar Beant Singh State University
Gurdaspur, Punjab, India

Gurpreet Singh
Centre for Biomedical Engineering
Indian Institute of Technology Delhi
New Delhi, India

ISSN 2524-5384 ISSN 2524-5392 (electronic)
Materials Horizons: From Nature to Nanomaterials
ISBN 978-981-99-5063-8 ISBN 978-981-99-5064-5 (eBook)
https://doi.org/10.1007/978-981-99-5064-5

This Springer imprint is published by the registered company Springer Nature Singapore Pte Ltd.
The registered company address is: 152 Beach Road, #21-01/04 Gateway East, Singapore 189721, Singapore

Preface

The book titled 'Materials for Biomedical Simulation: Design, Development and Characterization' is focused on the prospective materials for hard tissues, such as knee joints, hip joints, and bones, and soft tissues, such as skin, muscles, and functional organs. Comprising a total of eleven chapters under this title, this book covers a wide range of biomaterials simulating the properties of different hard tissues and soft tissues of the human body. These chapters emphasize the recent advances in biomedical material simulation and cover different materials for biomedical simulation, and focus on the designing, development and characterization of different implants, medical devices, and tissue surrogates.

Chapter 1 focused on the proficiency of auxetic metamaterials for their applications in biomedical and tissue engineering. Auxetic metamaterials can be used in a wide range of biomedical applications, such as cardiac stents, disks for spine support, hip implants, bone plates, nasopharyngeal swabs, etc. This chapter discussed the properties and major designs of auxetic materials. Nowadays, additive manufacturing is well established itself for developing complex auxetic geometries and structures in different applications, including biomedical engineering. Chapter 2 presented the recent advances in orthotic, prosthetic design with the associated challenges and applications. This chapter detailed the development procedure of prostheses and orthotics for patients and also compared their traditional manufacturing with additive manufacturing techniques. Chapter 3 elaborates on the current research and future trends of self-healing elastomer materials. This chapter discussed the processing and characterization of E-glass fiber-reinforced self-healing elastomeric composite materials and electrochemical discharge-based machining to drill the embedded HGTs. The investigation of self-healing elastomers is inspired by biological or natural systems in which damage prompts the healing process. Chapter 4 reported the biomechanical modeling of hierarchical metamaterials for skin grafting. This chapter used additive manufacturing to fabricate a hierarchy of Alternating Slit-shaped and Rotating Rectangle-shaped auxetic skin graft simulants on a skin simulant material to assess their expansion capabilities. The developed skin graft simulants were anticipated to be potentially applicable for skin grafting applications.

Chapter 5 reported a comparative study on the machining performance of cobalt-chromium and β-type titanium biomedical alloy. Both biomaterials were treated using electric discharge machining, and the modified surface was investigated for biomedical applications. Barefoot slips are most common in bathrooms, resulting in falls and related injuries, particularly in older adults. Chapter 6 investigated the traction performance of barefoot heel simulant, where different contaminated bathroom flooring tiles were used to assess the slips and falls. A human heel surrogate was prepared to study the slip safety of flooring tiles employed in the bathrooms and subjected to different bathroom contaminants. Such results with barefoot slip testing have not been reported to date and are anticipated to provide general guidelines for selecting suitable bathroom floorings.

Chapters 7 and 10 focused on developing and biomechanically testing biofidelic human soft tissue surrogates. Chapter 7 reported the fabrication and mechanical behavior of surrogates for the stomach tissue, whereas Chap. 10 discussed the development of tongue and tonsils tissue surrogates and, correspondingly, their biomechanically testing. Their linear and non-linear mechanical behavior was investigated, and different hyperelastic curve fit models were used to characterize and compare the results with literature studies. Such precisely characterized tissue surrogates with realistic mechanical properties are indispensable for studying various injury and disease biomechanics and modeling different medical models for trauma scenarios, surgical training, and educational purposes. Chapter 8 detailed the application of finite element analysis to estimate the mechanical performance of a hip joint prosthesis comprised of a Silicon Carbide fiber-reinforced Silicon Carbide matrix ceramic composite and Ti-6Al-4V. The prosthesis model was generated in SolidWorks and then imported into ANSYS for FEA analysis. The study's findings are promising, demonstrating that the use of Silicon Carbide fiber-reinforced Silicon Carbide matrix ceramic composite has the potential to revolutionize the hip joint prosthesis industry due to its remarkable mechanical properties and resistance to wear.

Chapter 9 overviewed the applications of nanomaterials in dental sciences and their scope in future practice. Nanoparticles provide unique mechanical, physical, and chemical properties to already existing dental materials based on polymers, resins, metals, and composites. This chapter will provide detailed information that nano dentistry can help in reducing the treatment time and improving the therapeutic outcome and maintenance in prosthodontics, endodontics, orthodontics, as well as other domains of dentistry. Bioprinting has emerged as a new technology for diagnostic research as well as the development of new organs or organics. In this direction, Chap. 11 discussed new biopolymers, including chitosan, collagen, hyaluronic acid, and silk fibroin, for their possible applications in bioprinting. This chapter determined the strengths and limitations of recently used polymeric biomaterials and 3D bioprinting technology.

This book is anticipated to serve as a key reference textbook for research in tissue engineering & biomedical engineering, biomaterials, biomechanics, and implant & medical device development with contributed chapters solicited in the areas of soft materials, such as elastomers, hydrogels, etc., for various applications; auxetic metamaterials; additive manufacturing of bio-implants; artificial tissues and organs; development of biomimetic materials; medical implants and biomedical device design; bioinspired and bio-tribological materials; advances in materials science for biomaterial applications; biomechanical characterization of hard and soft human tissues; bioprinting and nano-biomaterials.This book can be used as a comprehensive source for scientists, academics, researchers, and engineers in various areas and we are highly confident that this contribution will benefit all the readers in different ways.

New Delhi, India — Arnab Chanda
arnab.chanda@cbme.iitd.ac.in

Gurdaspur, India — Sarabjeet Singh Sidhu

New Delhi, India — Gurpreet Singh
July 2023
gurpreet.singh@cbme.iitd.ac.in

Contents

Editors and Contributors

About the Editors

Dr. Arnab Chanda works as Assistant Professor in the Centre for Biomedical Engineering, IIT Delhi, and Joint Faculty at the Department of Biomedical Engineering, AIIMS, Delhi. He is also the founder of a start-up company BIOFIT Technologies LLC, USA. Dr. Chanda is an expert in the fabrication and mechanical characterization of tissue mimics, and has previously developed artificial surrogates for human skin, muscles, brain, artery, and plantar fascia, and tested them in lab and clinical settings. These experimental models have been used extensively to conduct surgical training and study a wide range of injury scenarios. To date, he has received young researcher awards from ASME and MHRD and holds 7 US patents and 2 Indian patents. He has authored more than 85 articles in reputed international journals and several tech-transfers. Currently, Dr. Chanda heads the "Disease and Injury Mechanics Lab (DIML)," where his team works on developing cutting-edge healthcare technologies for disease mitigation for diabetic ulceration, cerebral aneurysm, severe skin burns, slips, and falls, etc., in India.

Dr. Sarabjeet Singh Sidhu is the Associate Professor and Dean Research at Sardar Beant Singh State University, Gurdaspur, Punjab, India. He completed his Ph.D. from Thapar University in 2014. Broadly, his area of research is related to surface modification, NDT, metallic composites, biomaterials, and non-conventional machining processes. He believes in collaborative work. As a result, he was invited to various international platforms as Joint Research Partner and Eminent Speaker. He has published 100+ technical papers in reputed national and international journals and conferences, acted as a potential reviewer for several journals, and chaired various technical sessions. His research is currently aimed at developing novel β-titanium alloys as biomaterials. He served as an editor of many reputed journals, books, and conference proceedings. Lately, he was awarded the ISTE Best Teacher Award 2021.

Gurpreet Singh is a Ph.D. scholar in the Centre for Biomedical Engineering, Indian Institute of Technology Delhi, India. He is a recipient of the most prestigious Ph.D. fellowship in India, the Prime Minister's Research Fellowship (PMRF), in the May 2021 cycle. His research interests include soft tissue mechanics, artificial tissues, biomimetics, and computational biomechanics. He is working on the computational modeling of diabetic foot and its urceleration progression. He is also working in developing artificial human tissue surrogates for injury and disease modeling. Previously, much of his work has been on improving the surface characteristics and bioactivity of metallic biomaterials, with research interests including surface engineering, materials science, biomaterials, and non-conventional machining processes. He worked on the surface modification of metallic biomaterials using electro-discharge machining, where he studied the bioactivity of modified surfaces in terms of wear resistance, corrosion resistance, and other biological responses. He has authored one book and contributed more than fifty research papers and book chapters to leading international journals and conferences. He is also serving as a reviewer for prominent journals worldwide.

Contributors

Mohammad Afazal University Polytechnic, Jamia Millia Islamia Delhi, New Delhi, India

Saba Afreen Indraprastha Dental College and Hospital, Ghaziabad, Uttar Pradesh, India

Vishnu Agarwal Department of Biotechnology, Motilal Nehru National Institute of Technology Allahabad, Prayagraj, India

Tushar Aggarwal Department of Mechanical Engineering, Manav Rachna University, Faridabad, Haryana, India

Jasdeep Bhinder École de Technologie Supérieure ÉTS, Montreal, Canada

Arnab Chanda Centre for Biomedical Engineering, Indian Institute of Technology (IIT), Delhi, India;
Department of Biomedical Engineering, All India Institute of Medical Sciences (AIIMS), Delhi, India

Subhodip Chatterjee Centre for Biomedical Engineering, Indian Institute of Technology Delhi, New Delhi, India

Sandeep Devgan Department of Mechanical Engineering, Khalsa College of Engineering and Technology, Amritsar, India

Shubham Gupta Centre for Biomedical Engineering, Indian Institute of Technology (IIT) Delhi, Delhi, India

Vivek Gupta Centre for Biomedical Engineering, Indian Institute of Technology (IIT) Delhi, New Delhi, India

Abhishek Kundu Department of Applied Mechanics, Motilal Nehru National Institute of Technology Allahabad, Prayagraj, India

Amit Mahajan Department of Mechanical Engineering, Khalsa College of Engineering and Technology, Amritsar, India

Vinod Mahajan Department of Mechanical Engineering, Khalsa College of Engineering and Technology, Amritsar, India

Biswarup Mukherjee Centre for Biomedical Engineering, Indian Institute of Technology Delhi, New Delhi, India;
Department of Biomedical Engineering, All Indian Institute of Medical Sciences Delhi, New Delhi, India

Viveksheel Rajput Sophisticated Analytical Instrumentation Facility (SAIF), Panjab University, Chandigarh, India

Neha Sardana Department of Metallurgical and Materials Engineering, Indian Institute of Technology Ropar, Rupnagar, India

Deepmala Sharma Department of Mathematics, National Institute of Technology Raipur, Chhattisgarh, India

Gaurav Pal Singh Department of Metallurgical and Materials Engineering, Indian Institute of Technology Ropar, Rupnagar, India

Gurpreet Singh Centre for Biomedical Engineering, Indian Institute of Technology (IIT), Delhi, India

Akhil Kumar Sonkar Department of Applied Mechanics, Motilal Nehru National Institute of Technology Allahabad, Prayagraj, India

Sahil Thakur Department of Mechanical Engineering, Manav Rachna University, Faridabad, Haryana, India

Anshul Tripathi Department of Mechanical Engineering, Manav Rachna University, Faridabad, Haryana, India

Chapter 1
Auxetic Materials for Biomedical and Tissue Engineering

Gaurav Pal Singh and Neha Sardana

1 Introduction

Auxetic metamaterials are a type of material that expands perpendicular to the direction of the applied force, as opposed to conventional materials that contract. Auxetic metamaterials are composed of a series of repeating subunits, or unit cells, that are arranged in a specific design to create the desired auxetic properties. The design and the material of the unit cells determine the overall properties of the metamaterial. Auxetic materials offer a unique set of properties that have the potential to revolutionize materials science and engineering. They are being actively studied and developed by researchers around the world with the aim of creating new materials that are stronger, lighter, and more versatile than conventional materials.

Auxetic materials display a negative Poisson's ratio, which is rarely found naturally and for very low strains. It has been reported that cat skin [1] and cancellous bone [2] show auxetic behavior, and some of the human body parts such as annulus fibrous of the intervertebral disk [3], arteries [4], skin tissues [5], and ligament tissues [6] also show some degree of auxetic behavior under certain conditions. Healthcare is an essential aspect of modern society and plays a vital role in promoting the overall health and well-being of individuals and communities. Human body parts and organs get damaged or dysfunctional due to old age, accidents, or diseases. The replacement of these organs is generally provided by organ donors. But the demand for compatible organs is generally very difficult to fulfill. Tissue engineering is an interdisciplinary field that combines biology, engineering, and materials science to create biological substitutes that can restore or replace tissues and organs which are damaged or diseased. Tissue engineering has the potential to revolutionize the healthcare industry by offering new approaches to treating injuries and diseases that

G. P. Singh · N. Sardana (✉)
Department of Metallurgical and Materials Engineering, Indian Institute of Technology Ropar, Rupnagar 140001, India
e-mail: nsardana@iitrpr.ac.in

A. Chanda et al. (eds.), *Materials for Biomedical Simulation*, Materials Horizons: From Nature to Nanomaterials, https://doi.org/10.1007/978-981-99-5064-5_1

currently have limited or no effective treatments. Thereby improving the quality and lifespan of humans and other living beings. Tissue engineering is not only being explored to build new organs but also for the safe testing of drugs without using humans or animals test subjects. Furthermore, in depth disease studies can be performed by creating models of cells such as cancer or tumors, which is often unpredictable and laborious while using living subjects. The critical aspect of tissue engineering is the scaffold which creates the base upon which cells of the tissue can be grown in three dimensional (3D) space [7].

The conventional methods of scaffold development are fiber bonding, freeze drying, leaching, and gas foaming. However, it is excruciatingly difficult to control the porosity and the pore interaction of the structures fabricated by these methods. Moreover, the response of these structures does not accurately represent the actual tissue. The advancement of 3D printing and other fabrication techniques has allowed researchers to use auxetic materials as scaffolding structures. The auxetic materials are also able to closely mimic the slight negative Poisson's ratio behavior shown by the human tissues. Along with tissue engineering, Auxetic materials are used to create healthcare devices such as implants, stents, and cardiac patches. These materials can improve the mechanical properties of the implants and reduce the risk of implant failure. Also, the auxetic materials are lighter and stronger than the conventionally used materials.

The current chapter aims to introduce auxetic metamaterials along with their properties, commonly used design configurations, and fabrication methods. The applications of metamaterials in the field of biomedical simulation will be discussed to present the current state of the art technology in auxetic materials.

2 Properties of Auxetic Materials

2.1 Negative Poisson's Ratio

Poisson's ratio (ν) is a measure of how a material deforms (expansion or contraction) perpendicular to the direction of loading. It is defined as the negative ratio of the transverse strain (ε_t) to the longitudinal strain (ε_l) [8]:

$$\nu = -\frac{\varepsilon_t}{\varepsilon_l} \tag{1}$$

The Poisson's ratio is between 0 and 0.5 for most of the commonly used materials. Auxetic materials are unique because they exhibit a negative Poisson's ratio, meaning that they expand laterally when stretched longitudinally, rather than contracting as most materials do, as shown in Fig. 1a. The negative Poisson's ratio is a fundamental characteristic of auxetic materials. By expanding laterally when stretched, auxetic materials can offer several advantages over traditional materials, such as increased

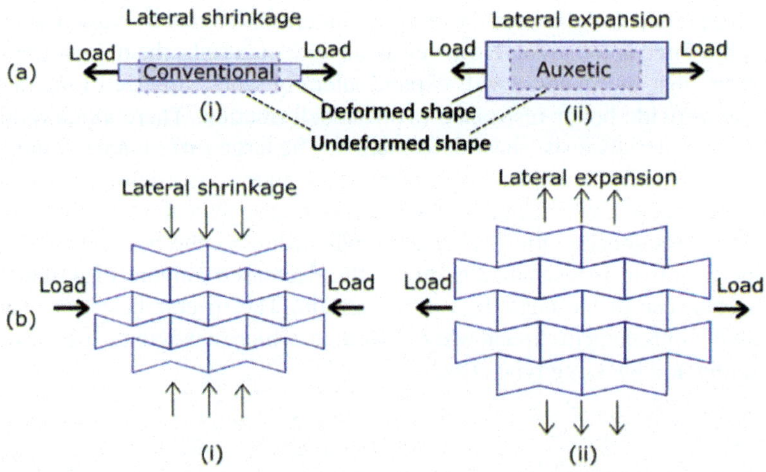

Fig. 1 Schematic of comparison of a **a** conventional and auxetic material. **b** Auxetic effect using hexagonal re-entrant structure under tensile and compressive loading. Reproduced from [9], CC BY 4.0

flexibility and better shock absorption. Additionally, they can provide better resistance to certain types of deformation and failure, such as wrinkling and buckling, which can be important in a variety of engineering applications. The Poisson's ratio of auxetic materials can vary depending on the specific structure and composition of the material, as well as the loading conditions. In general, auxetic materials typically have a Poisson's ratio in the range of −1 to −0.5. A common configuration to achieve the auxetic behavior is shown in Fig. 1b.

2.2 *Shear Resistance*

The shear resistance of auxetic materials is very different from traditional materials due to their negative Poisson's ratio. The relation between shear modulus (G) and Poisson's ratio (ν), bulk modulus (K), and Young's modulus (E) for isotropic solids is [10, 11]:

$$G = \frac{3K(1 - 2\nu)}{2(1 + \nu)} \tag{2}$$

$$G = \frac{E}{2(1 + \nu)} \tag{3}$$

From Eqs. 2 and 3, it can be observed that when Poisson's ratio tends to −1, the shear modulus approaches infinity. Thereby, auxetic materials tend to have a high shear modulus. In traditional materials, when a shear force is applied, the material tends to deform in a particular direction, which can cause the material to fail if the force is too great. However, in auxetic materials, the negative Poisson's ratio means that the material will expand laterally when stretched longitudinally, which can provide better resistance to shear deformation. There are several ways that auxetic materials resist shear deformation. The large surface area of the auxetic materials increases the contact points that are available between different components of the material, which enables the material to distribute the forces more evenly. Thus, the stress concentration is reduced, which lowers the possibility of failure. Also, increased shear resistance is due to the decreased amount of wrinkling and buckling because of the negative Poisson's ratio. The inverse behavior of auxetic materials enables them to expand when deformed, which makes them less susceptible to wrinkling and buckling type effects.

2.3 *Indentation Resistance*

The indentation resistance is the ability of a material to resist deformation when a load is applied to it. The negative Poisson's ratio enables the auxetic material to resist deformation because the material flows into the region where the load is applied, as shown in Fig. 2. The inverse is true when a conventional material undergoes the same load. Classically, in the elastic region, the indentation resistance can be approximated as hardness (H) is related to Poisson's ratio (ν) and Young's modulus (E) as [12].

$$H \propto \left[\frac{E}{(1-\nu^2)}\right]^{\gamma} \tag{4}$$

where γ is a constant related to the direction of the load applied. It can be concluded from Eq. 4 that the hardness of the material (or resistance to indentation) approaches infinity for Poisson's ratio close to 1 or −1 [14]. Furthermore, the periodic structure

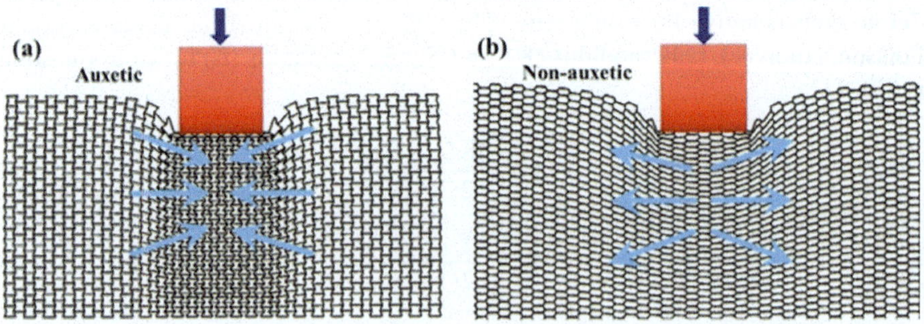

Fig. 2 Schematic representation of indentation resistance mechanism of an **a** auxetic and **b** non-auxetic material. Reproduced from [13], © 2021 with permission from Elsevier Ltd

of auxetic materials provides additional structural stability and rigidity by spreading the load across a larger volume.

2.4 Fracture Resistance

Similar to shear and indentation resistance, auxetic materials display a higher fracture resistance than traditional materials [15, 16]. The tendency of the auxetic material to expand upon stretching prevent cracks from spreading through the material and can also help to absorb and distribute energy from an applied load or impact. Yang et al. [17] reported that auxetic composites showed a fracture toughness improvement of more than two times than the non-auxetic composite with similar materials. The relation of the critical tensile strength (σ_c) with the Poisson's ratio (ν), Young's modulus (E) is [18]:

$$\sigma_c = \left[\frac{\pi ET}{2r(1-\nu^2)}\right]^{1/2} \quad (5)$$

where T is the solid surface tension, and r is the radius of the plane circular crack. As evident by Eq. 5, the material becomes infinitely tough as Poisson's ratio approaches 1 or -1.

2.5 Synclastic Behavior

Synclastic behavior is the property of the material to curve in the same direction for longitudinal as well as transverse deformation when a load is applied upon it. The synclastic behavior is important in applications such as the design of thin-walled structures and biomedical implants. When an out of plane bending moment is applied, traditional materials show a saddle shape, as shown in Fig. 3b, and auxetic materials form a dome shape, as shown in Fig. 3a. The synclastic behavior is a convenient method to produce dome type shapes without using complex and expensive fabrication methods [8].

2.6 Variable Permeability/Porosity

Variable permeability is important in applications such as filtration systems, drug delivery devices, and tissue engineering. Auxetic materials display a load dependent permeability/ porosity due to the orientation of the unit cells. The representation of the variable permeability to the induced strain in the auxetic material is shown in Fig. 4.

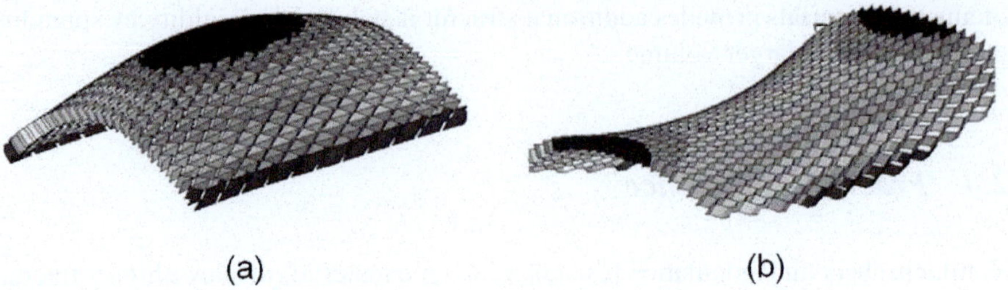

Fig. 3 Schematic representation of out of plane bending of an **a** auxetic and **b** non-auxetic material. Reproduced from [19], © 2010 with permission from Elsevier Ltd

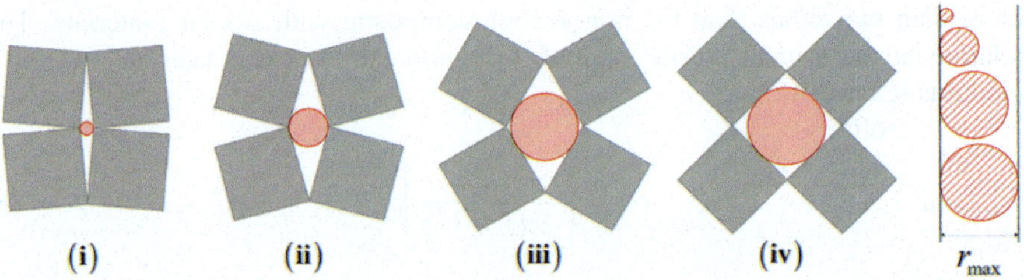

Fig. 4 Representation of the increasing permeability from (i) to (iv) of an auxetic material based on the rotating square's design. Reproduced from [21], CC BY 4.0

Moreover, the load can be mechanical, thermal, chemical, or even electromagnetic in nature. This property is essential to regulate or control the flow of fluids or gases in filtration systems that use auxetic materials. One potential application of variable permeability auxetic materials is in drug delivery systems. The permeability of the material is varied in response to a specific stimulus, so it is possible to control the rate and extent of drug release, improving the efficacy and safety of the treatment [20].

2.7 *Energy Absorption*

The unique deformation behavior of auxetic materials enables increased energy absorption. Further, the unit cells can be filled with appropriate materials, which can enhance the energy absorption ability of the auxetic materials. This property has many applications, including impact protection, automotive safety, and sports equipment. Imbalzano et al. [9] fabricated an auxetic composite panel that could absorb twice the energy compared to the monolithic panel when subjected to blast loading. Also, the velocity of the back panel was reduced by 70%. The unique structure dependent properties of auxetic materials enable controlled energy transfer to the desired areas. Commercially, auxetic materials are being used in shoe soles to reduce the impact on the foot as well as spread the energy to locations that minimize the risk of injury.

3 Types of Auxetic Structures

The discovery of the auxetic behavior of materials was made more than a century ago [22]. However, the experimental validation of their existence was confirmed in 1944 in a research article by Love [23], in which iron pyrite crystal was reported to have a negative Poisson's ratio. In 1998 Baughman et al. [32] determined that auxetic behavior is shown by 69% of cubic metals consisting of a single element when deformed in [110] direction. Recently, first principle calculations and molecular dynamics have enabled researchers to investigate the auxetic nature of various materials. Molecular dynamics simulations were used to establish the auxetic nature of ideal zeolitic structures for a particular direction of deformation [25, 26]. However, only a few experimental studies confirm the simulation results [27]. First principle calculations were used to establish that 2D materials, including δ phosphorene [28], MoC_2 [29], graphene [30], and TiN [31], show auxetic behavior at the nanoscale. The negative Poisson's ratio of the 2D materials makes them an excellent choice for nanodevice and nanoelectronic applications. Organic materials such as membranes of red blood cells [32] and nucleus of embryonic stem cells [33] have been also reported to show auxetic properties in the microscale. As mentioned in the introduction, skin and ligament tissues and fibers of intervertebral disk and arteries also show some auxetic behavior.

Although auxetic behavior is sparingly found in nature on the macro scale, however, it is activated only in a small range of strains and/or for particular loading directions. Thereby, their utilization for commercial applications is not feasible. The first man made material that was designed to show auxetic properties was fabricated by Lakes in 1987 [18]. Since then, research groups from all over the world have contributed to improving the performance of auxetic materials. There are various design strategies for fabrication that are applied as per the properties required of the auxetic materials. The important structures are discussed in this section.

3.1 Re-entrant Structures

The term re-entrant is defined as a structure that is directed inward [34]. In the context of auxetic materials, the simplest structures which are considered re-entrant are similar to a bow tie, as shown in Fig. 5a. When the structure is deformed in the longitudinal direction, the overall system expands in the transverse direction (instead of the expected collapsing) due to the opening up of the bow tie shape.

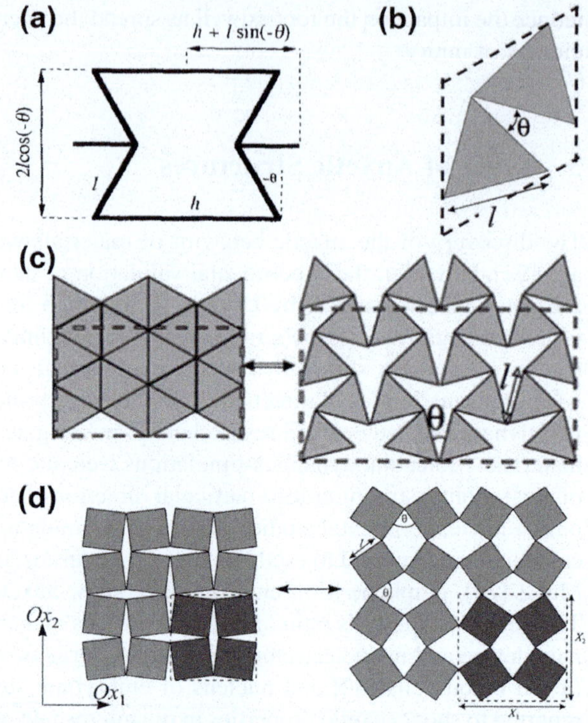

Fig. 5 **a** Bow tie type re-entrant geometry. Reproduced from [35], © 1996 with permission from Elsevier Ltd. **b** Repeating unit cell and **c** geometry of rotating triangles type structure. Reproduced from [36], © 2006 with permission from Springer nature. **d** Geometry of rotating squares type structure. Reproduced from [37], © 2000 with permission from Springer nature

3.1.1 Foams

In the first reported auxetic material fabricated in a laboratory, Lakes [18] developed an open cell foam in which the ribs of the cell were directed inwards to form a re-entrant type structure. The material used was polyester, and the Poisson's ratio achieved was −0.7. The foam was prepared by compressing a conventional polyester foam triaxially and heating it above the softening temperature. The foam was compressed, and it achieved a permanent re-entrant structure. It was concluded that the auxetic foam does not conserve volume, and it was more resilient than the conventional foam. Further, synclastic behavior was also observed in the fabricated auxetic foam.

Generally, auxetic foams have higher shear modulus, toughness, and energy absorption as compared to conventional foams [38, 39]. But their Young's modulus is lowered due to the ease of deformation as a result of the presence of buckled ribs in the foam structure [40]. The mechanism of the unique properties of auxetic foams has not been established by one theory by the scientific community. Gibson and Ashby [41] used 2D hexagonal honeycombs to model the conventional foams. When the auxetic behavior was introduced into the foams, Choi and Lakes [42] reported a method that added the effect of rotation caused by plastic hinge formation and was based on the strain energy technique. Grima et al. [43] presented a method that assumed that the hinges and joints were constrained to move and most of the deformation was at the ribs. A combination of mechanisms is at play in the re-entrant structure that induces the auxetic properties in the foams.

Auxetic metallic foams have been used in stents and bioimplants because they can reduce the risk of failure by distributing the stresses during impact. Metallic foams can be fabricated through powder metallurgy, casting, electrodeposition, solid state processing, and additive manufacturing. However, the working strain in which the auxetic nature is active is very low for metallic foams as compared to the polymer based foams. The polymer based foams developed by the thermal process have the problem that the auxetic structure slowly returns to the conventional form due to creep [44]. Scarpa et al. [45] reported that a change in volumetric compression of up to 30% could be observed a week after the auxetic foam was prepared. The optimization of the temperature and lubrication of the molds was considered a possible solution [44]. Other solutions which redesigned the heating processes include multi-stage heating [44] and solid state foaming [46]. However, the thermo-mechanical method of fabrication is laborious and, therefore, not suited for mass production. Hence, researchers have developed chemical routes to develop the re-entrant structures in conventional foams.

3.1.2 2D Structures (Honeycombs)

As explained previously, the typical hexagonal structure of a re-entrant structure has a bow tie type shape which induces auxetic properties in one direction of loading; however, the overall system is highly anisotropic [35]. As compared to the conventional honeycomb structure, the bow tie shape results in higher transverse Young's and shear modulus [47]. The Poisson's ratio (ν) and Young's modulus (E) of the bow tie type hexagonal structure in the loading direction are:

$$\nu = \frac{\sin\theta(h/l + \sin\theta)}{\cos^2\theta} \tag{6}$$

$$E = k\frac{(h/l + \sin\theta)}{b\cos^3\theta} \tag{7}$$

where the symbols h, l, b, and θ are defined in Fig. 5a. The parameter $k = E_s b(t/l)^3$ and E_s is the intrinsic Young's modulus of the material of which the hexagonal walls are constructed. The transverse shear modulus was reported to increase as the rib slenderness ratio (ratio of the width to the length of the rib of the hexagon) decreased [48]. The finite element method (FEM) was applied to understand the effect of the geometry of the structures on the overall properties of the material. It was observed by Whitty et al. [49] that the rib thickness had a massive effect on the stiffness, Poisson's ratio and Young's modulus. It was concluded that if the thickness of the diagonal ribs was similar to or less than the vertical ribs, the flexing of diagonal ribs had a major contribution to the overall deformation of the system. However, when the thickness of the vertical ribs was lower, the stretching of vertical ribs dominates the deformation when loaded in the y direction. Similarly, many other groups used FEM simulations to understand the deformation mechanism of the hexagonal structures and optimize the geometry to achieve the desired properties [50, 51]. Bezazi et al. [51] modified the hexagonal geometry by adding base walls to reduce stress concentration effects by reducing the sharp edges. The design was deemed to be more manufacturable as compared to the ideal structures; however, a lesser negative Poisson's ratio was observed. The design was further modified by altering the geometry of the base wall and adding a narrow rib in the structure. It was observed that the Poisson's ratio could be tuned by varying the geometric and material parameters. The aforementioned studies were valid for small deformations and in the elastic range. Modified material models for large deformations were developed by Wan et al. [50] and Yang et al. [52].

Besides the honeycomb structure, other structures such as arrow [53], lozenge [54], and star shaped [55] were also developed to achieve auxetic properties. Under compression, the triangles present in the arrow shaped structures will close to produce a transverse contraction. The Poisson's ratio achieved by Larsen et al. [53] was −0.8, which was lower than most of the hexagonal honeycomb structures. The star shaped structures fabricated by Grima et al. [55] extended the concept of honeycomb structures by connecting 3, 4, and 6 sided stars to achieve rotational symmetry. The 3 sided star structure was observed to display auxetic and conventional behavior depending upon the force constants that were induced. The 4 and 6 sided star systems exhibited negative Poisson's ratio for most of the conditions. Gaspar et al. [54] fabricated Lozenge and square grid structures to achieve Poisson's ratio of −0.43 and −0.6, respectively. Elipe et al. [56] compared the various 2D honeycomb structures using FEM simulations. For the re-entrant type structures, the square grid structure achieved the most negative Poisson's ratio of −0.9. In summary , sinusoidal, triangular, 4 star, and Lozenge grid displayed Poisson's ratio of −0.81, −0.79, −0.50, and −0.33 respectively.

3.1.3 3D Structures

Although auxetic foams offer good auxetic properties, their variability and inability to finely tune the properties based on the loading direction inspired researchers to look into other methods. Therefore, building upon the 2D re-entrant structures, 3D structures were introduced to achieve the aforementioned missing properties. Furthermore, the 3D structure and the potential to control the auxetic properties in all three major directions increases their applicability. The advancement in additive manufacturing techniques provided access to quickly and easily build complex 3D structures. Evans et al. [57] in 1994 presented on of the first 3D re-entrant structures which used a dodecahedron geometry. It was shown that a negative Poisson's ratio was possible in all three directions. The geometry of the structure was analyzed using FEM, and rib flexing was deemed to be the dominant mechanism of deformation. Wang et al. [58] constructed and studied a cylindrical 3D structure that constituted re-entrant

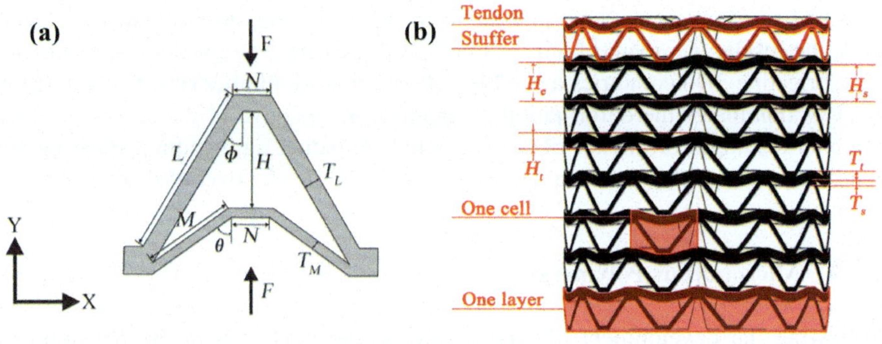

Fig. 6 **a** Unit cell and **b** layers of the 3D cylindrical auxetic material. Reproduced from [58], © 2016 with permission from IOP Publishing Ltd

triangles, as shown in Fig. 6. The effective Young's modulus and Poisson's ratio in the y direction (in the direction of the height of the cylinder) was derived as:

$$E_y = \frac{HKE\alpha^3}{2K^2\beta\sin^2\Phi + (K + K^2)\sin^3\Phi} \tag{8}$$

$$\nu_y = \frac{\left(H - \frac{H}{KL}\right)\sin\Phi\cos\Phi + \left(\beta H - \frac{\beta H}{L}\right)\cos\Phi - \frac{H^2}{L}\sin\Phi - \frac{\beta H^2}{L}}{\left(L - \frac{L}{K}\right)\sin^3\Phi + \left(3\beta L - \frac{\beta L}{K}\right)\sin^2\Phi + 2\beta^2 L\sin\Phi} \tag{9}$$

where Φ, N, M, H, and L are shown in Fig. 6a and K = M/L and β = N/L. E is Young's modulus of the material, and α is the rib slenderness ratio. As evident from Eq. (8), as the Young's modulus is increased, Poisson's ratio decreases with the angle Φ, which is the angle between the wall and y axis. The tuning of the overall properties of the auxetic material can be accomplished by optimizing the geometry of the unit cell. Li et al. [59] used a hollow skeleton to construct the re-entrant structure; it was observed that the auxetic nature of the overall structure increased as the re-entrant angle increased. Yang et al. [60] used electron beam lithography to fabricate a 3D re-entrant structure which displayed a higher compressive strength than auxetic foams. Also, the Ti–6Al–4 V-based auxetic material had higher strength and stiffness along with a larger negative Poisson's ratio. Other additive manufacturing techniques such as laser based optical lithography [61], inkjet printing [62], dual material polyjet method [63], and 3D printing [64] have been used by researchers to fabricate 3D re-entrant auxetic structures. Yang et al. [65] studied the mechanism of 3D auxetic structures for small as well as large strains under the assumption that the axial shrinkage of the ribs of the structure due to axial stress was not significant. Thereby, the study successfully modeled the auxetic material when the ribs were slender, but the calculations on the thicker ribs did not yield satisfying results. Wang et al. [66] used the energy method to model the 3D re-entrant type structure. The effect of rib slenderness ratio and angle of the re-entrant structure was analyzed, also, the results were valid both for thin and thick ribs. The results were validated by fabricating 3D geometries available in literature. For thinner ribs, the bending of the ribs dominates the deformation mechanism irrespective of the direction of the uniaxial loading. For thicker ribs, various factors including, bending, shearing and loading direction play a role in defining the behavior of the overall structure.

3.1.4 Microporous Polymers

Following the development of auxetic foams, the next step in the development of auxetic materials was that of microporous polymers by Caddok et al. in 1989 [67]. Following the sintering of polytetrafluoroethylene, it was rapidly heated and expanded by drawing. The resulting polymer was highly porous on the micron level and had a maximum negative Poisson's ratio of -12. However, it was reported that the deformation mechanism of the disk shaped particles and fibrils made the material highly anisotropic. When the material was not under any loading, the material was highly dense. As a tensile force was applied, the fibrils stretched and caused the particles to rotate, and an expansion was observed. Eventually, particles reached their final state, and maximum expansion was observed at low strains [68]. At higher strains, the stiffness increased and the Poisson's ratio became less negative as the particles rotated beyond the optimal position. Alderson and Evans [69] used a 2D model to understand the behavior of a microporous polymer. The model assumed that the material comprised of rectangular nodes which were connected by fibrils, similar to the system explained by Caddok et al. [67]. The equations for Young's modulus and Poisson's ratio were derived and validated using experimental results. It was reported that the material properties and geometry of the fibrils are essential to calculate the force coefficients for the deformation. Rigorous testing of the characterization of the material was performed to estimate the auxetic behavior of polytetrafluoroethylene microporous polymers. Pickles et al. [70] used thermoforming to fabricate a microporous polymer using ultra high molecular weight polyethylene (UHMWPE). The powder of the polymer was compacted, sintered, and then extruded through a die to obtain the microporous structure. The same method was applied to the polymer polypropylene also; however, the negative Poisson's ratio obtained was relatively small. Alderson et al. [71] compared the properties of auxetic, compression molded, and sintered UHMWPE. It was reported that for loads up to 100 N, the auxetic effect was preserved. The auxetic material displayed a higher indentation resistance as compared to conventional polymers. Further, the auxetic polymer showed the highest plasticity and had the fastest viscoelastic response. Although the microporous polymers display a large negative Poisson's ratio, easy fabrication, and stable auxetic structure, their use in commercial applications has been limited due to their highly anisotropic nature and low reproducibility of properties.

3.2 Rotating (Semi-) Rigid Structures

Auxetic structures can be constructed using squares, triangles, parallelograms, and other shapes which can rotate over hinges to expand/contract depending on the type of loading which is applied. Grima [36] modeled the rotating triangles type mechanism based on the assumptions that the triangles were non-deformable in the loading directions and there was no shear in the system. The triangles rotated to create the auxetic effect. The equations for the Young's modulus, Poisson's ratio, and compliance matrix (S) of the system were derived using the conservation of energy model:

$$E_1 = E_2 = K\frac{4\sqrt{3}}{l^2\left[1 + \cos\left(\frac{\pi}{3} + \theta\right)\right]} \tag{10}$$

$$\nu_{12} = {\nu_{21}}^{-1} = -1 \tag{11}$$

$$S = \begin{pmatrix} S_{11} & S_{12} & 0 \\ S_{21} & S_{22} & 0 \\ 0 & 0 & 0 \end{pmatrix} = \frac{1}{E}\begin{pmatrix} 1 & -1 & 0 \\ -1 & 1 & 0 \\ 0 & 0 & 0 \end{pmatrix} \tag{12}$$

where K is the stiffness constant of the hinges about which the triangles rotate, θ is the angle between two triangles as shown in Fig. 5b, l is the length of the triangle, and E is the inherent Young's modulus of the material. Similar to the triangular structures, Grima et al. [37] also derived the equations for the aforementioned parameters of rotating rigid squares under the same assumptions. The equations are as follows:

$$E_1 = E_2 = K\frac{8}{l^2[1 - \sin\theta]} \tag{13}$$

$$\nu_{12} = \nu_{21} = -1 \tag{14}$$

$$S = \begin{pmatrix} S_{11} & S_{12} & 0 \\ S_{21} & S_{22} & 0 \\ 0 & 0 & 0 \end{pmatrix} = \frac{1}{E}\begin{pmatrix} 1 & 1 & 0 \\ 1 & 1 & 0 \\ 0 & 0 & 0 \end{pmatrix} \tag{15}$$

where K is the stiffness constant of the hinges about which the squares rotate, θ is the angle between two squares, as shown in Fig. 5c, and l is the length of the square. Assuming that the constituent structures were semi-rigid, the rigidity of the structures, hinges, and loading direction all have a major effect on the Poisson's ratio of the system [37]. The rotating rectangles added more complexity to the deformation mechanism; the Poisson's ratio could be positive or negative depending on the angle between the rectangles [72]. The dimensions of the rectangles also play a vital role in determining the Poisson's ratio of the auxetic material. Moreover, the Young's modulus was dependent on the angle between rectangles. As the angle increased, the theoretical modulus increased and approached infinity and then decreased. Although the mathematical models and equations provide a good estimate of the Poisson's ratio and Young's modulus, the idealistic assumptions lead to some variations as compared to the experimental values. Mathematical models often overestimate the auxetic behavior of the materials. Grima et al. [73] improved the semi-rigid model by considering the square rotating structures could be deformed under axial and shear loading such that the sides of the squares could be alternated to create rectangular structures. The new model created was able to predict the properties of zeolite crystals; however, a slight overestimation of auxetic behavior was still observed. Shan et al. [74] concluded that the isotropy of the rotating structures based on auxetic materials was dependent on the degree of rotational symmetry of the system. Attard et al. [75] extended the concept of rotating structures into 3D using a cuboidal shaped network. It was shown that the structure could exhibit auxetic behavior in all three directions. Apart from triangles and squares discussed in this section, structures comprising of rhombi and parallelograms were analyzed by research groups. Rhombi have two types of designs (type α and type β) in which the unit cell can be constructed. In the type α configuration, the obtuse angle of the rhombus is connected to the acute angle of the next rhombus. In the type β configuration, the acute angle is connected to the acute angle (obtuse is connected to obtuse) of the next rhombus [76]. It was reported by Attard et al. [76] that both configurations were able to display auxetic nature. The type α configuration's auxetic nature was highly angle dependent, and the response was anisotropic. The type β configuration displayed in-plane isotropy, and the Poisson's ratio was close -1. Further, the Poisson's ratio did not have any dependence on the acute or obtuse angles of the rhombus. The parallelogram based structures display 4 types of configurations in which the parallelograms can be connected [77]. As observed in the rhombi, varying nature of auxetic behavior is shown by each of the configurations.

3.3 Crumpled Sheets

The inspiration for planar or crumpled auxetic sheets was to improve reproducibility and reduce the anisotropy in auxetic foams. Alderson et al. [78] uniaxially compressed a polyurethane foam to 40% to 60% of its thickness to produce a flat sheet of thickness less than 3 mm. A Poisson's ratio of -3 and -0.3 were observed for flat and curved sheets, respectively. The crumpling of the pores in the sheet at the microscopic level was the major cause of the auxetic behavior. Although the flat sheets produced a lower Poisson's ratio, their results were much more consistent than the curved sheets.

The aforementioned sheet displayed auxetic response at the micro level. For some materials, the auxetic behavior is even observed at the nano scale. Ma et al. [79] produced a crumpled nanopaper by aerosolization and quickly drying a graphene oxide solution. Hall et al. [80] calculated the Poisson's ratio of single and multi-wall carbon nano tube (CNT) sheets (buckypaper) using an optical microscope. The response of the buckypaper was recorded as it was deformed at a constant rate. It was observed as the concentration of multiwall CNTs was increased, the Poisson's ratio went from positive value to negative. Moreover, significant changes in other properties, including Young's modulus, toughness, strength, and electrical conductivity, were also observed. It was reported that increasing the diameter of the multiwall CNTs with a large number of interior walls led to a large negative Poisson's ratio. Grima et al. [81] observed auxetic behavior in corrugated graphene monolayers in specific directions. The properties of the corrugated graphene sheets were dependent on the geometry of the applied corrugation as well as introducing defects into the graphene sheets. Molecular dynamics was used to model and calculate the properties of the graphene sheets. The same group also reported the auxetic properties of crumbled graphene sheets [82].

Another inspiration to study crumpled sheet type auxetic materials is the art of paper folding called Kirigami. One of the simplest methods of using Kirigami to produce auxetic materials is cutting repeating designs of V shaped cuts arranged in a specific way. When the material was stretched, the V shaped cuts open up, leading the material to expand. Eidini et al. [83]created a unit cell which consisted of two zig zag strips in which a parallelogram shaped hole was cut out. The Miura-ori based pattern showed tunable auxetic behavior depending upon the location of the zig zag strips. Liu et al. [84] studied the response of the sheets fabricated using Miura-ori method. Experimental and FEM simulations were performed to understand their response. Elvaloy AC 1820 polymer was selected to fabricate the patterned sheet. The pellets of the polymer were melted and formed into a sheet which was later compressed to for the desired pattern upon it. The thickness of the sheet was approximately 1.8 mm. Out of plane compression test, in-plane compression test, and three point bending test were performed on the patterned sheet. The out plane tests initiated buckling of the connecting faces followed by collapse and densification. The in-plane tests resulted in lower bending and displayed higher compression ratios. The sheet was developed for applications cushioning in sports shoes and other wearable devices.

3.4 *Perforated Sheets*

Perforated sheets are flat materials which have regular pattered holes. As opposed to the hinges present in the rotating structures, the perforated sheets show auxetic behavior by the expansion/compression of the holes in the sheets. Perforated sheets are generally stiffer and stronger than rotating structures, although rotating structures posses higher flexibility, hence, a larger range of Poisson's ratio. Grima et al. [85] created diamond (Fig. 7a) and star shaped holes in the commonly available material used in carpets to display the auxetic behavior of perforated sheets. Theoretically, the Poisson's ratio of the perforated sheet with diamond shaped holes can be calculated as:

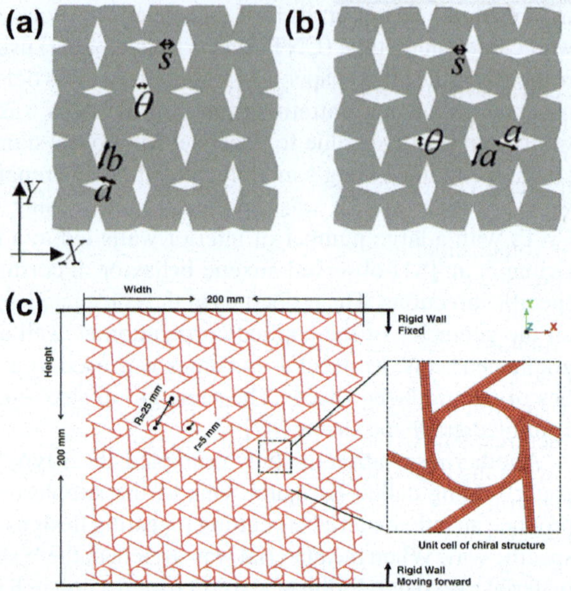

Fig. 7 Perforated sheet type auxetic material with diamond shaped perforations with sides of **a** different and **b** same sizes. Reproduced from [85], © 2010 with permission from John Wiley and Sons. **c** Auxetic material with chiral geometry. Reproduced from [86], © 2021 with permission from Elsevier Ltd

$$\nu_{xy} = {\nu_{yx}}^{-1} = \frac{a^2\cos^2\left(\frac{\theta}{2}\right) - b^2\sin^2\left(\frac{\theta}{2}\right)}{a^2\sin^2\left(\frac{\theta}{2}\right) - b^2\cos^2\left(\frac{\theta}{2}\right)} \tag{16}$$

where a and b represent the sides of the rectangle and θ is the angle between two consecutive rectangles, as shown in Fig. 7a. The experimental results were supported by FEM simulations in ANSYS software. It was observed that at small strains, a rotating structures type mechanism was observed, but as the deformation was increased to larger values, the mechanism became insignificant. It was also reported that as the material in the gaps between the structures increased, the system tended to lose its auxetic nature. The deformation of the shape of the structures was deemed to be the major cause of the aforementioned phenomenon. Mizzi et al. [87] analyzed an auxetic material by creating slit patterned holes. The highest Poisson's ratio of the system was −13. Grima et al. [88] created a perforated material which comprised of randomized cuts in the sheet. The FEM simulations, as well as experimental results, confirmed that the auxetic nature of the overall material was present even when slits were present at random locations in the sheet. Thereby establishing that a highly symmetric structure is not necessary to achieve auxetic behavior. Carta et al. [89] created patterned slit perforations in a polycarbonate sheet. The design was envisioned to decrease stress concentration in the material. An isotropic auxetic behavior was observed as a result of the 45° rotation of the perforated slits. The results were analyzed experimentally using the Digital Image Correlation (DIC) method and were supported by FEM calculations.

3.5 Chiral Structures

A chiral structure refers to an object or structure which cannot be exactly aligned with its mirror image. The auxetic structures created can be chiral or anti-chiral. Anti-chiral structures display reflective symmetry. A chiral unit cell is shown in Fig. 7c. A central cylinder is tangentially connected to ribs which are in turn further connected to their respective cylinders [90]. When a load is applied, the cylinder will rotate, creating flexing in the ligaments, which folds or unfolds them depending upon the type of loading. Generally, isometric auxetic chiral structures can achieve Poisson's ratio close to −1. Pure chiral structures have to adhere to the limitations of rotational symmetry. Thus, the number of connecting ligaments have to be equal to the order of the symmetry [91]. However, chiral auxetics which do not obey the laws of rotational symmetry can also be developed. Contrary to the re-entrant type structures, the Poisson's ratio of the overall structure is not dependent on the angle between the ligaments. Further, the range of working strain is higher than re-entrant type structures [92]. The deformation is dependent upon how the ligaments rotate/wind onto the cylinders. An increase in the number of ligaments leads to higher stiffness of the structures. Alderson et al. [93] reported that chiral honeycombs displayed a higher Young's modulus than the anti-chiral honeycomb having an equal number of ligaments. The three main chiral structures which obey rotational symmetry are tri (3 ligaments), tetra (4 ligaments), and hexa (6 ligaments) chiral structures. Alderson et al. [93] showed that anti-trichiral achieved a small negative Poisson's ratio of −0.11 when the ligament length was small. Poisson's ratio was 0.08 when the ligament length was large due to the flexure of the ligaments. The auxetic behavior of the structures was found to increase when tetrachiral structures were analyzed [93]. The chiral as well as anti-chiral structures achieved Poisson's ratio of −1. Mousanezhad et al. [94] reported that the Poisson's ratio of the chiral structures varied significantly when the length of the ligaments was varied in the x and y direction. But an equal ligament length produced an isotropic auxetic material with Poisson's ratio −1. A hexa-chiral structure shows a hexagonal symmetry because each cylinder is connected to 6 ligaments. The structures are generally isotropic and have Poisson's ratio near −1 [92, 93]. Gao et al. [86] studied the high strain impact performance of chiral auxetic structures fabricated by water jet machining of an aluminum alloy panel. The chiral unit cell consisted of 4 cylinders, and each cylinder was connected to 6 ligaments. The high strain rate testing was performed in the Split Hopkinson pressure bar (SHPB) apparatus, and the deformation and strains were mapped using the DIC technique which uses high speed cameras to observe the changes in the sample as the material is subjected to loading. Further, the experimental results were validated using FEM simulations in the ABAQUS software. It was reported that the failure of the structure when subjected to high strain rate loading was initiated in three modes, buckling of the ligaments, buckling of the cylinder and ligaments, and crushing of the cylinders and ligaments, when the impact velocity in the SHPB apparatus was 5, 25, and 50 m/s respectively. The energy absorption of the material increased as the relative density of the structure was increased.

3.6 *Auxetic Gels*

Auxetic gels contain a network of interconnected pores or channels that elongate when the material is stretched, and an overall expansion is observed in the material. Auxetic gels have major applications in tissue engineering and soft robotics. The gels are usually polymer based and have a large surface area, high porosity, and excellent absorptivity. Their properties can be tuned to replicate biological tissues; however, auxetic gels have been difficult to fabricate [95]. Ma et al. [96] fabricated a polyvinyl alcohol (PVA) hydrogel with internal by-concave pores, which were interconnected to each other. The PVA powder was converted to PVA hydrosol using an electro thermic pressure steam chamber. A surfactant was added to create pores in the hydrosol, and the obtained solution was frozen for a few hours. The samples were defrosted and then compressed triaxially to 5% of the initial volume and then again frozen. A freeze thaw cycle was performed six times, and the obtained sample was washed in water using an ultrasonic cell pulverizer. The sample obtained contained internal by-concave pores. For the creation of interconnected pores, NaCl particles were mixed with the surfactant in a 1:1 ratio which was added to the PVA hydrosol. The remaining steps were as that of the previously mentioned process, except that the compression steps were not performed. The maximum negative Poisson's ratio observed was −0.83 and −1 for internal by-concave pores and interconnected pores, respectively. Bhullar et al. [95] added acrylamide hydrogel to an auxetic polytetrafluoroethylene jacket. It was reported that the swelling ratio of the jacket was dependent on the amount of hydrogel added. The addition of hydrogels improved the shear strength, Young's modulus, and fracture toughness of the hybrid auxetic material while achieving a Poisson's ratio of −0.98. It was concluded that the hybrid auxetic material displayed high compatibility with biological materials and had improved mechanical properties. Kunwar et al. [97] used optical projection lithography to fabricate a double network hydrogel which was structured to form a re-entrant pattern. A bow tie shaped honeycomb geometry was constructed out of a polymer solution. The resulting structure combined the benefits of the auxetic properties of a re-entrant structure with the biocompatibility of hydrogel type material.

4 Fabrication of Auxetic Biomaterials

The advancements in fabrication methods have enabled researchers and manufacturers to produce complex structures with advanced materials. Simple fabrication methods such as casting are not suitable for fabricating complicated 3D patterns. Techniques such as laser cutting [88, 98] and lithography [95] are effective in creating 2D structures. However, the most widely used method to produce 3D auxetic materials currently is additive manufacturing or 3D printing. The simplicity of fabricating 3D structures at the macro or micro scale and the flexibility of using materials ranging from polymers to metals make additive manufacturing such an attractive fabrication method. The general steps involved in additive manufacturing are making the computer aided design (CAD) of the desired structure, selecting the materials, and then printing the geometry layer by layer to generate the structure. The cost of additive manufacturing setup depends on the materials being used and the required resolution of the parts. Hence, commercial 3D printers are available to develop low cost prototypes as well as highly sophisticated finished products. The manufacturing industries have now begun to embrace 3D printers to produce finished commercial products. The commonly used additive manufacturing techniques which are used to fabricate auxetic materials are Stereolithography (SLA), Digital Light Processing (DLP), Fused Deposition Modeling (FDM), Melt Electrowriting (MEW), and Multiphoton Lithography (MPL).

SLA method uses a laser to cure a material (generally liquid) to the shape of the desired pattern. SLA was instrumental in the development of 3D auxetic materials. The pioneering research by Maruo et al. [99] in 1986 used a UV laser to cure liquid monomers, which were deposited layer by layer via a syringe to fabricate a complex 3D auxetic material. The selection of the laser is determined by the chemical response of the monomer. Further, post printing steps, including cleaning and additional curing, are generally required to improve the properties of the final product. Warner et al. [100] developed an auxetic material to build scaffolds for muscle and tendon tissues. An advanced SLA based printer that used UV light was selected to cure isobornyl acrylate solution. The total time required to print the multiyear scaffold was around 3 min. The auxetic material which was printed contained sharp and rounded hinges in a honeycomb type pattern. The solution was structured to the required geometry and then cured by the UV laser. The remaining monomer, which was not exposed to the UV light was removed by dipping the whole structure in chloroform. The scaffolds were aimed to assist the cell growth and to facilitate tendon to tissue interface. The tendon tissue connection is critical to prevent injuries such as tendinopathy, tendon rupture, and other degenerative disorders. Wagner et al. [101] fabricated a re-entrant type auxetic material that was used for temperature dependent structure transformations. The advantages of SLA method include a smooth surface finish that does not require much finishing, the speed of the process, and its good accuracy. The resolution which can be achieved by SLA printing is of the order of hundreds of microns.

The DLP method improves the SLA method by introducing sub-micron resolution and flexibility to print various materials, including polymers, resins, and even some metals. The DLP method is even faster and supports continuous printing of multilayer structures due to the efficient curing or the photopolymerization step [102]. The liquid polymer is exposed to the selected light from a projector through an arrangement of lenses. The polymer is cured layer by layer, starting from the top and moving downwards (as opposed to the traditional 3D printing methods) [103]. Wu et al. [104] fabricated an auxetic material for use in endovascular stents using the DLP method. The major polymers used were poly(ethylene glycol) diacrylate oligomers, butyl methacrylate, and butyl acrylate. A 385 nm UV light source was used to cure the liquid. The thickness of the layers that were printed was 50 μm, and the time required for printing each layer was less than 10 s. Li et al. [105] developed a re-entrant type auxetic material using a state of the art printer, nanoArch P140. A high power UV source of wavelength 405 nm was used, and the resolution in the x and y directions was less than 10 μm.

The FDM method is the most commonly used 3D printing technique in which a heated nozzle is used to melt and extrude a thermoplastic filament. The polymer is deposited layer by layer on a stage to create the required structures. The filament is generally fed through a spool to the nozzle to create a continuous process. The materials which can be printed using the FDM method are acrylonitrile butadiene styrene (ABS), Polylactic acid (PLA), nylon, and many more. Yang et al. [106] fabricated a re-entrant type hexagonal shaped honeycomb using PLA and polyurethane (PU) polymers. A Poisson's ratio of −1.82 was achieved for the re-entrant type structure. It was reported that the auxetic material had increased shock absorption, which could potentially reduce the risk of injury as compared to its non-auxetic variant. Although the FDM technique is economical and easy to use, its limited material compatibility and lower resolution (in the millimeter range) as compared to other additive manufacturing methods.

The SLS and selective laser melting (SLM) methods (both works on similar principles) use powdered materials which are sintered by a laser to build a layer of the required pattern. Generally, a new layer has to spread on the platform to build the next layer. The biggest advantage of the SLS method is the variety of materials that can be used, plastics, ceramics, and metals can be fabricated. The resolution which can be achieved is higher than the FDM method, structures containing features of a few hundred microns can be printed using the SLS method [107]. The energy incident on the powders should be enough to attain a phase transition from solid to liquid or semiliquid form. The temperature of the chamber is below the melting point of the material, and the small spot size of the laser beam helps ensure a local rise in temperature. Kapnisi et al. [108] developed an auxetic material with a honeycomb shaped re-entrant type auxetic material for the treatment of myocardial infarction. A 248 nm pulsed KrF laser was incident on a chitosan polyaniline composite. The repetition rate was 40 Hz and the pulse energy was 350 mJ/cm^2. The re-entrant type structure was covered with polyaniline to make the structure electrically conductive. The auxetic material displayed a maximum negative Poisson's ratio of −1.45 was achieved. Gang et al. [109] developed stents of various designs using polyamide powder. The SLS method was used to fabricate the stents with varying geometric parameters to optimize the design of the unit cell. The stents had good mechanical properties, and the values of the negative Poisson's ratio achieved were between −0.3 to −2.3. Recently, Meng et al. [110] presented a biodegradable auxetic material for the development of scaffolds in soft tissues. Polycaprolactone polymer was structured using the SLS method in a pattern where the unit cell was shaped with sinusoidal elements. A layer of powder with a size of 100 μm was rolled over on a platform, and it was sintered using a laser of power 3 W with a spot size of about 150 μm. After the solidification of one layer, the platform was moved down by the same height as that of the layer (0.15 mm), and this process was repeated until the desired structure was fabricated.

MEW and melt electrospinning writing (MES) methods are used to fabricate micro and nano structures using a nozzle from which an electric field is used to draw out and elongate a polymer jet. The polymer is melted and then extruded using a high voltage electric field. Next, the polymer is deposited on a substrate layer by layer to create the desired structure. The MEW method is increasingly being used to develop scaffolds for tissue engineering. This method is able to create complex pore structures with interconnectivity. Recently, Paxton et al. [111] developed an auxetic material that was used to create scaffolds. The MEW method was used to fabricate the re-entrant tubular type structure out of polycaprolactone polymer. The polymer was heated and then extruded out of a syringe nozzle. A voltage of 5.5 kV was applied between the nozzle and a rotating substrate. The pattern was created by printing at a constant rate of 100 mm/min. Olvera et al. [112] fabricated a cardiac patch using an auxetic based device with a missing rib type structure. Polycaprolactone was fabricated using the MEW method which was melted at 85 °C and then extruded using an acceleration voltage of 6 kV at a pressure of 0.22 bar. Fibers created were of diameter of approximately 120 μm and the velocity attained was 6 mm/s. To create the electrical conductivity of the cardiac patch, a coating of polypyrrole polymer was used. The conductive polymer had the potential for applications in printed electronics as well as biomedical devices.

The MPL is a technique that uses a laser to solidify a photosensitive resin. The laser beam of a very small spot size and high intensity can create patterns with a resolution of a few hundred nanometers. The resin is in the liquid state, and complex optics are used to create a 3D printed sample layer by layer. The resin which is not solidified acts as a support for the remaining structure. The printed sample is washed with a solvent to remove the unreacted resin and then cured to improve the mechanical properties of the structure. Flamourakis et al. [113] created bow tie shaped re-entrant type auxetic scaffolds using the MPL technique. The photosensitive organic-inorganic SZ2080 polymer was drop cast onto glass substrate which was coated with a monomer to promote adhesion. 800 nm femtosecond pulsed Ti Sapphire laser was incident on the sample. A piezoelectric stage, along with a 100X objective lens was used to create the desired pattern onto the polymer.

Apart from additive manufacturing, other fabrication techniques are required to develop certain auxetic materials. As explained in Sect. 3.1.1 auxetic foams are created by the stress compression method in which the foam is prepared by compressing a polymer triaxially and heating it above its softening temperature. The polymer based foams developed by the thermal process have the problem that the auxetic structure slowly returns to the conventional form. The thermo-mechanical method of fabrication is laborious and therefore not suited for mass production. Hence, researchers developed chemical routes to develop auxetic foams. Grima et al. [114] placed the compressed foam in acetone for about an hour and cured it at room temperature to achieve a re-entrant structure. The properties obtained were very similar when compared to foams fabricated by thermo-mechanical processes. Another advantage of using the chemical route is that in some cases, the auxetic foams can be converted back to conventional foams by the same chemical which was used to fabricate them. The chemical used to achieve the re-entrant structure may also be in a gaseous form which also simplifies the fabrication process. The chemical route can also be used to fabricate structures in nanowires [115] and nanoflakes [116] in polymers. However, the nanowires are unstructured and are distributed randomly in the matrix material.

5 Applications of Auxetic Materials in Biomedical Engineering

5.1 Implants

Severe cases of lower back pain as a result of bulging of the spinal cord due to injury or disease are treated by replacing the intervertebral disk with an implant. The properties of the implant are a critical factor in ensuring the complete recovery of the patient. Martz et al. [117] theoretically analyzed an auxetic material that was designed to be used as an intervertebral disk to prevent the nerve impinging due to undesired bulging. The mechanical behavior of the auxetic implant was similar to that of the natural body part, which ensured that undesired stresses due to the implant were low. A bow tie type geometry was used to create the implant geometry, and FEM simulations were performed to understand the response of the implant. The Poisson's ratio of structured high density polyethylene was confirmed experimentally was performing displacement controlled measurements. The experimental results were in close agreement with the FEM simulation. Jiang et al. [118] also created an intervertebral disk implant with an auxetic structure, as shown in Fig. 8a. Polyurethane was patterned to create a Bucklicrystal by 3D printing via the SLS method. Compression tests were performed to ensure the stability of the auxetic structure. Various mechanical tests were performed using FEM simulations. The stress contours showed that the auxetic material displayed better stress distribution, as shown in Fig. 8b. Both in vitro and in vivo studies were performed the study the real life performance of the implant. The in vitro studies were performed for one week in a simulated environment. The in vivo studies involved performing disk replacement surgeries on 20 New Zealand white rabbits. The Magnetic resonance imaging (MRI) tests were performed after 2, 4, and 8 weeks of surgery. The implant showed excellent biocompatibility and was able to withstand practical loading conditions. It was reported that the use of auxetic material increased the energy absorption of the implant when under sudden loading.

Pedicle screws are used in spinal surgery to stabilize and support the spine. The screws ensure that a strong foundation is formed upon which the fusion surgery used to join two or more vertebrae together can be performed. They are also used to treat spinal deformities and stabilize the spine after a fracture or dislocation. Yuan et al. [119] used auxetic structures to fabricate pedicle screws to improve bone screw contact. A re-entrant type geometry was created by 3D printing Ti6Al4V using the SLM method. It was reported that the properties of the screw could be tuned

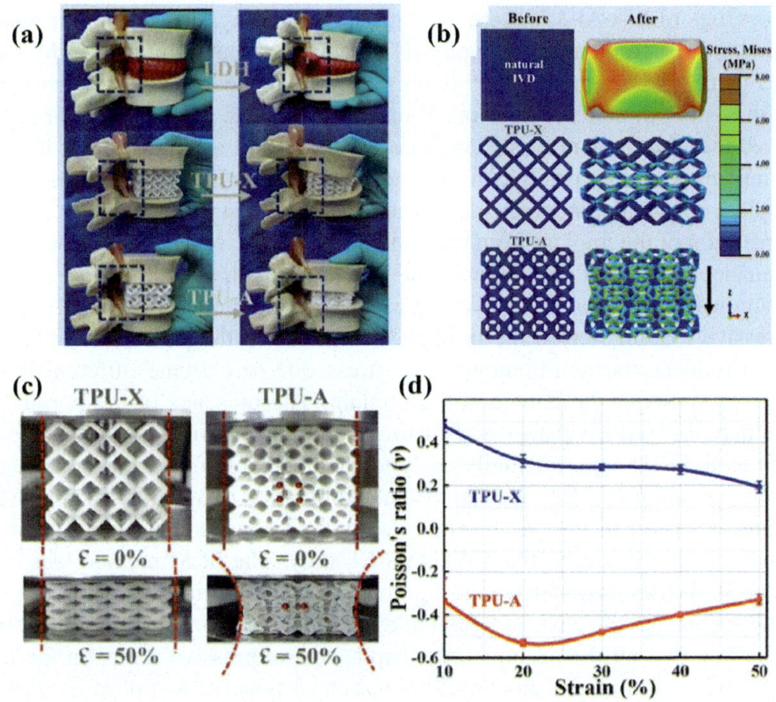

Fig. 8 **a** Lumbar disk herniation model of empty, non-auxetic implant (TPU-X) and auxetic implant (TPU-A). **b** Stress distribution under compressive loading of Intervertebral disk (IVD), TPU-X and TPU-A. **c** Experimental images and **d** Poisson ratio of TPU-X and TPU-A under compressive loading. Reproduced from [118], CC BY 4.0

by varying the re-entrant angle from 45° to 90° and the thickness of the ribs. An experimental setup was created to replicate the condition in which the screw was pulled out of the bone. The screw was fixed into a synthetic bone block, and a load cell was attached to the screw to apply the pulling force. A load of 10kN was applied on the screw, and a displacement speed of 5 mm/min was applied according to the standard testing procedure. The expansion of the auxetic material of the screw ensured the screw did not protrude out of the indented location when the screw was under tensile loading. The stress analysis was confirmed by performing FEM simulations.

Knee replacement implants are increasingly becoming more familiar due to the rise of total knee replacements to treat the pain caused by arthritis and other chronic pain and mobility problems. The knee joint undergoes heavy loading cycles while walking and while performing most physical activities. Hence, the current generation of implants have limited life due to the wear of the implant and the material surrounding the implant. Ghavidelnia et al. [120] theoretically analyzed four types of knee implants, solid, structured with positive Poisson's ratio, structured with negative Poisson's ratio, and a hybrid with a combination of positive and negative values of Poisson's ratio. The geometry of the structure was a 3D re-entrant type, and the properties of Ti-6Al-4 V were used to model the implant. The femur bone was designed using computed topography (CT) scan of a healthy human in MATLAB. The geometry was imported to Solidworks software to clean out the geometry, which aided in a smooth and accurate mesh in the simulation software ANSYS. It was observed that the implant was in both compressive and tensile loading; hence, the knee implant with only positive or only negative Poisson's ratio leaves some part of the implant constantly under tensile loading. The constant tensile loading can lead to the separation of the implant from the bone, which can have a negative impact on the performance and life of the implant. The hybrid implant, which had Poisson's ratio distribution from 0.3 to −0.3, overcame the aforementioned problem by creating compressive loading across the implant. Furthermore, the porous structure of the implants reduces the non-homogeneous stress due to extreme differences in the Young's modulus of the bone and the implant. The presence of the porous structure reduces this sudden change and ensures a gradual variation in Young's modulus. Kolken et al. [121] experimentally designed and tested hip implants comprising of conventional, auxetic materials, and also a combination of auxetic and conventional materials as shown in Fig. 9a, b. The bow tie type honeycomb structure was used to impart auxetic nature to the Ti6Al4V-ELI alloy. The SLM method was used to fabricate all the designs, which were rigorously tested in simulated conditions which faithfully represent the environmental and loading conditions in the human body. It was reported that all the designs only displayed compressive stains on the medial side. The hybrid implant (using a combination of positive and negative Poisson's ratio) also provided compression at the lateral implant bone implant interface. The purely auxetic implants did not provide the required compression due to the undesired deformation caused by the increased shear forces. The compression forces are usually considered less detrimental as compared to tension when the part is under shear loading [122]. The improved bone implant contact lowers the risk of wear and increases the longevity of the implant. Furthermore, the porous structure creates enhancement in bone growth as the joint underdoes loading cycles [123].

5.2 Stents

Stents are small devices that are placed in tubes or openings of the body that help expand and open up obstructions [124]. Oesophageal cancer is one of the major causes of malignant cancer deaths. Ali et al. [125] fabricated an auxetic stent and its grafts which help in the treatment of oesophageal cancer by reducing tumor growth, and it can also help prevent dysphagia. A sheet containing the rotating squares was constructed with polyurethane material using various methods, which included laser cutting followed by computer numerical control (CNC) machining and polymer casting by using a mold of the reverse structure fabricated using electron beam melting (EBM). The stent was fabricated by casting method in which molds were made using FDM and EBM. It was reported that laser cutting provided an economical and accurate method to produce the auxetic material. Vacuum casting

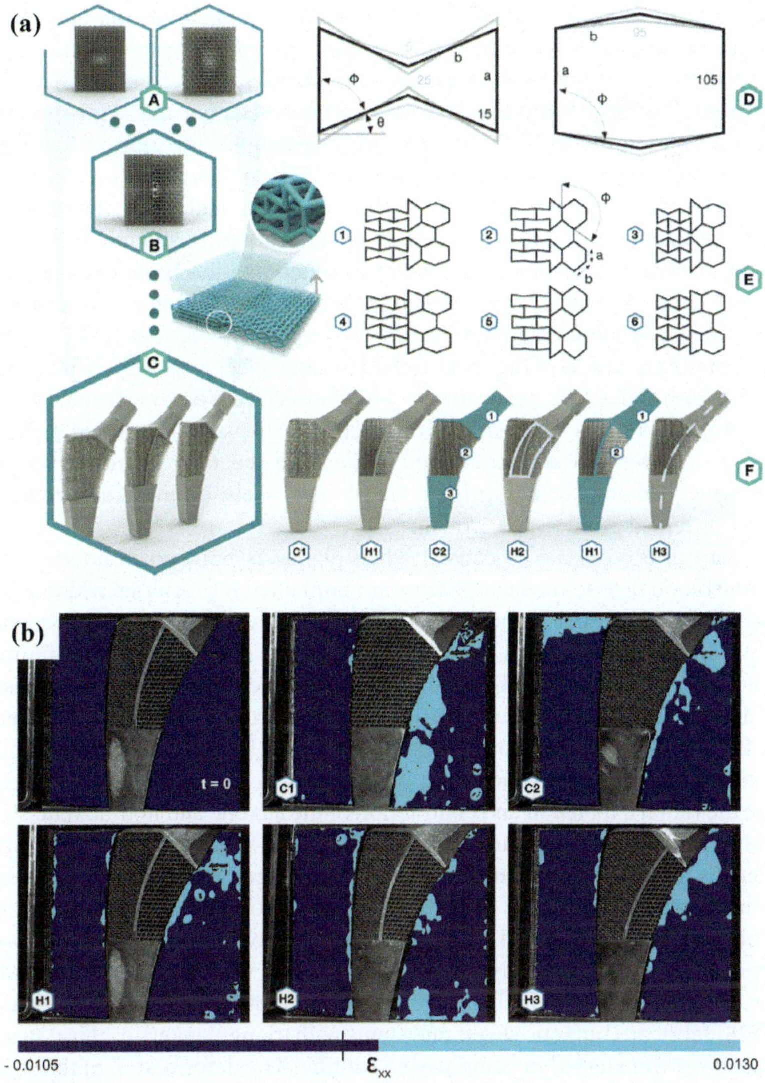

Fig. 9 Schematic illustration of the designs of the hip implants, **a** (A) Auxetic and non-auxetic structures, **b** hybrid structure, (C) and hybrid implant. (F 1–6) Various auxetic designs and (F) Non-auxetic implant (C1), Auxetic implant (C2), Hybrid implants (H1-3). **b** Horizontal strain response of the various hip implants at time 0 s and 180 s at 1.5 mm displacement. Reproduced from [121], © 2014 with permission from Royal Society of Chemistry

was determined to be the superior casting method. Tensile testing and expansion tests were performed to establish the mechanical properties and failure limits of the auxetic stents. The Poisson's ratio of the stents ranged from −0.87 to −0.96 for different loading conditions. It was also observed that the semi-rigid auxetic stents expand more, both in the radial and longitudinal direction, as compared to rigid material. Bhullar et al. [126] used the MEW technique to create microfiber sheets out of polycaprolactone. The microfiber sheets and solid polycaprolactone sheets were then further press molded to create a rotating squares type structure. Polycaprolactone was selected because of its biocompatibility and low degradation speed. The final stent shape was created by rolling the sheet in a cylindrical shape, and the joint was soldered together. Tensile test results showed that the Poisson's ratio and the Young's modulus of the solid and microfiber sheet were 0.55 N/mm and −1.1 and 1.75 N/mm and −1.02, respectively. Simulations were also performed to confirm the experimental results. The microfiber sheets also displayed increased stiffness and higher elastic deformation as compared to the solid sheets. It was observed that the auxetic stents had an excellent grip on the blood vessels, offered unrestricted blood flow, and provided good tissue growth.

Coronary artery disease is caused when plaque is built up in arteries that can reduce the blood flow to the heart. Stents are implanted into arteries which expand to ensure that there is proper blood flow to the heart. Lee et al. [127] developed vascular grafts which had the potential to be used to treat a variety of conditions, including coronary artery disease, peripheral artery disease, aneurysms, and vascular trauma. A cut missing rib type structure fabricated using projection based 3D printing in which UV light cured the photosensitive monomer (SLA method). Poisson's ratio of the auxetic material was measured by performing strain tests. The pulling tests confirm the radial expansion and contraction of the auxetic and non-auxetic stents, as shown in Fig. 10a, b. Cell cultures were performed to study the in vitro performance of the grafts. It was observed that the sample with negative Poisson's ratio displayed higher interconnectivity of the cell growth. It was reported that after one day, similar cell proliferation was observed for the negative and positive Poisson's ratio samples. However, beyond the first day, higher cell density was seen in the sample with negative Poisson's ratio, which can be clearly seen in Fig. 10 (c and d). Hamzehei et al. [128] fabricated auxetic sheets and stents with triangular anti-trichiral structures. The structures were tested in compressive loading. The deformation of the structures was due to the rotational movements of the triangles and the bending of the ribs which connected the triangles. The geometric parameters of the structures were varied to control the properties of the auxetic material. Thermoplastic polyurethane elastomer was selected to fabricate the auxetic sheets. FDM type 3D printing was used to fabricate the auxetic sheets. Laser cutting was performed on stainless steel to produce the cylindrical stents. Two types of triangular geometries were studied; the most negative Poisson's ratio achieved by each of the optimized geometries was −1.6 and −2.19. Conventional anti-trichiral stent geometry was also produced using stainless steel to compare the performance of the triangular auxetic stents. It was reported that lateral displacement and energy absorption of triangular geometry was much higher than the traditional anti-trichiral structures. Lin et al. [129] fabricated an

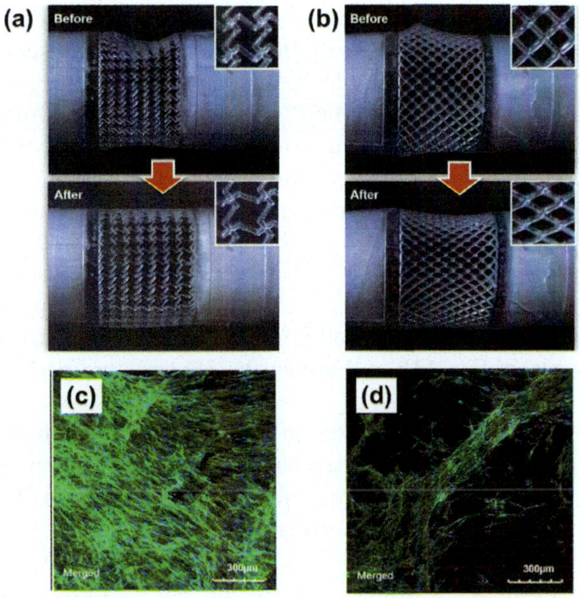

Fig. 10 Optical microscopy images of the pulling test of **a** auxetic and **b** non-auxetic stent. Stained actin filaments and nuclei after one week of cell culture for **c** auxetic and **d** non-auxetic stent. Reproduced from [127], CC BY 4.0

auxetic stent with shape memory properties. The shape memory behavior provides an additional parameter for the researchers to control the shape, size, and properties of the auxetic material. Generally, a temperature based phase transformation creates reversible expansion or contraction in the shape memory material. The shape memory in stents enables the material to expand or contract according to the temperature of the blood flowing across it. Shape memory polylactic acid (PLA) polymer was 3D printed based on the FDM technique. Stents of two configurations, both based on the re-entrant type structure were tested. It was reported that the recovery ratio for both configurations was more than 90% for temperature change between 70 °C and room temperature. In vitro experiments were performed in sheep intestines, both the configurations expanded within 5 s. The research confirms the feasibility of auxetic shape memory stents in healthcare applications.

5.3 Tissue Engineering

The properties of auxetic materials can be tuned by using specific materials and structures to resemble human tissues. Hence, auxetic materials have a massive potential for applications across tissue engineering.

Flamourakis et al. [113] created 3D auxetic scaffolds based on re-entrant geometry using the MPL method as shown in Fig. 11a. The material used was a hybrid polymer containing both organic and inorganic components called SZ2080. The mechanical properties of the bow tie shaped auxetic structure were measured on the nano scale.

Scaffolds based on two sizes of unit cells were studied, 8.6 μm (small pore size) and 40 μm (large pore size). In situ images were observed for the mechanical testing, and it was reported that the auxetic structures were able to compress uniformly and without cracking until 60% deformation. Cell culture studies were also performed for 3D as well as 2D auxetic scaffolds. It was reported that the small pore size did not allow any of the cells to penetrate its structure. However, these scaffolds were useful in creating good contact between the fibers and scaffolds and allows the fibers to grow and stabilize on the scaffold. The porous nature of the auxetic structure allowed the hard SZ2080 polymer to be tuned to achieve the desired properties. The large pore size structures allowed cell penetration and after 4 days of cell culture, there was excellent cell growth on the scaffold structure. Both phenomena can be seen in Fig. 11b, c. Further, the 3D structure performed much better than the 2D structure.

Chen et al. [130] tested auxetic materials, which were of the perforated sheet configuration and made using a cell laden fish gelatin methacrylamide hydrogel. The structuring of the hydrogel was performed using an FDM based 3D printer. The perforations in the patten were 2.9 mm in length and 0.3 mm wide, as shown in Fig. 12a.

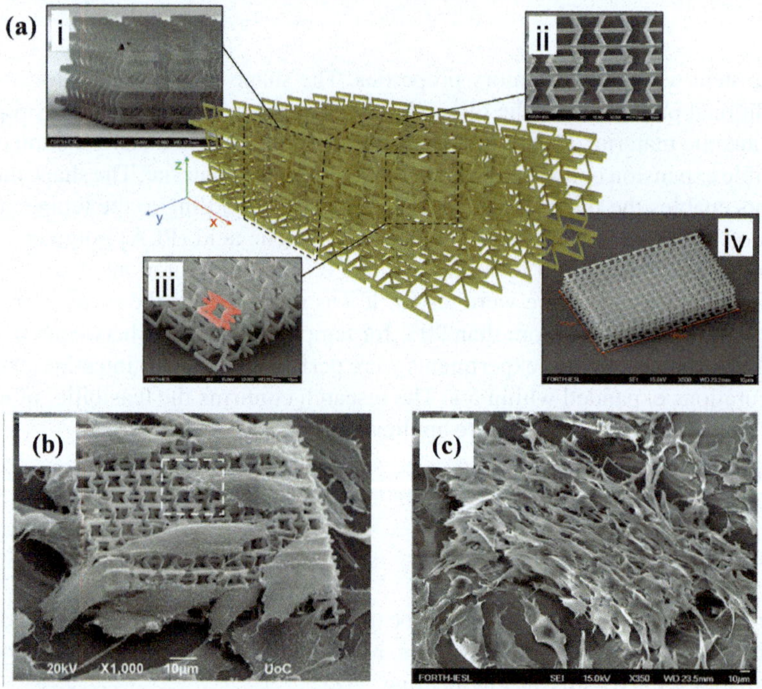

Fig. 11 Scanning electron microscopy (SEM) images of the auxetic scaffold with small pore size in the (i) tiled view, (ii) top view, (iii) magnified view, and (iv) tiled view. Cell culture images of the auxetic scaffold with **c** small and **d** large pore size. Reproduced from [113], © 2020 with permission from John Wiley and Sons

The human Schwann cells were cultured on the auxetic scaffold for one week. It was reported that the scaffold was able to endure up to 20% tensile strain. The cells showed excellent growth and spread across the scaffold structure. The scaffold with the Schwann cells was also tested when under tensile strain loading, an increased nerve growth factor and its receptor (tropomyosin receptor kinase A (TRKA)) was observed as shown in Fig. 12c. TRKs are a family of receptors which play an important role in the growth, development, and survival of nerve cells. Warner et al. [100] developed auxetic scaffolds to facilitate tissue regeneration. Two types of honeycomb geometries were selected, one with acute hinges and the other with rounded hinges. STL based 3D printing was used to fabricate the polymer (polyaliphatic urethane acrylate blend with isobornyl acrylate) into their respective auxetic structures. The material was deemed more elastic than most of the commonly used polymers and could withstand more strain as compared to most hydrogels. The mechanical testing of both types of structures was used to optimize the design and create the stabilized rounded hinge type geometry. In vitro experiments were performed using fibroblast and myoblast cells to evaluate the performance of the auxetic scaffolds. It was reported that the optimized design was able to enable cell proliferation within the voids, which was not observed with the traditional honeycomb type structures. Sonam et al. [131] controlled the Poisson's ratio of scaffolds to optimize their structure for tissue engineering applications. The STL method was used to create single as well as multilayer structures of polyethylene glycol-based material. The geometry of the structure was designed to create positive and negative Poisson's ratio at specific locations of the material. Cell culture studies were performed for one week, and the scaffold was seeded with human mesenchymal cell stem cells. The cell growth was observed over the whole scaffold structure. In the regions where negative Poisson's ratio was present, the cell growth was displayed both over the struts and in the voids. While in the positive Poisson's ratio regions, growth was not seen in the voids. The performance of the hybrid auxetic material was much better than the conventional scaffolds. The developed auxetic scaffolds had the potential for ulcer treatment due to the similarity of their properties with the desired tissue.

A cardiac patch refers to a small piece of tissue engineered material that is used to repair or replace damaged cardiac tissue. Kapnisi et al. [108] fabricated a conductive cardiac patch with a honeycomb re-entrant structure for the treatment of myocardial infarction using an SLM type 3D printing. The auxetic structure was covered with polyaniline to make the material electrically conductive. The properties of the auxetic patch were comparable to that of the heart tissue. Ex vivo experiments revealed that the auxetic cardiac patch was able to replicate the movements of the heart tissue, which was not observed for the unstructured cardiac patch. Two week in vivo experiments with the rat model of myocardial infarction confirmed the practical feasibility of the auxetic cardiac patch. Olvera et al. [112] developed an auxetic cardiac patch using a missing rib type structure. Polycaprolactone was fabricated using the MEW method. The auxetic structure was coated with polypyrrole polymer to create electrical conductivity. Cell culture study was performed by seeding the auxetic cardiac patch with bone marrow mesenchymal stromal cells. The cell proliferation was higher for the auxetic cardiac patch as compared to the simple square structured cardiac

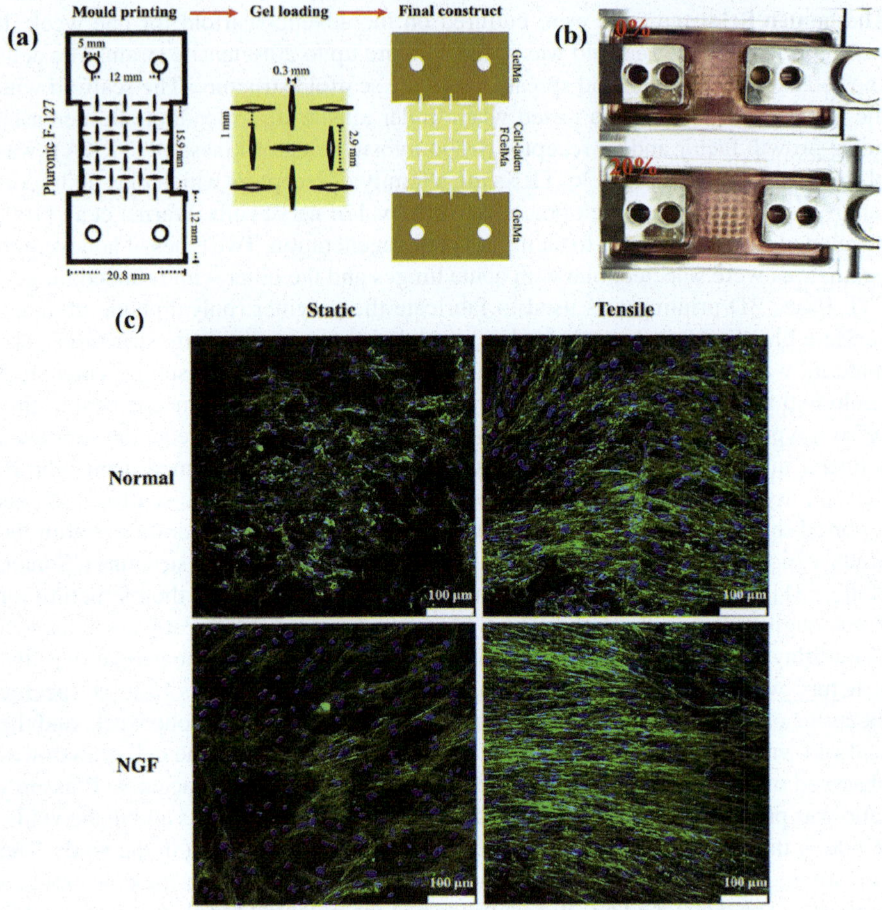

Fig. 12 **a** Schematic representation of the fabrication of perforated sheet hydrogel based auxetic material. **b** Cell culture experiments under tensile loading at 20% strain at a frequency of 0.48 Hz. **c** Fluorescent staining of the cells cultured with and without nerve growth factor, and under static and dynamic tensile loading. Reproduced from [130], © 2020 with permission from Elsevier Ltd

patch. The auxetic cardiac patch was stretched up to 20% strain, and the shape was recovered completely when under repetitive loading cycles. Dynamic loading did not alter the electrical conductivity of the auxetic structure. The cardiac auxetic patch was compatible with the human mesenchymal stromal cells and could withstand the deformation and stress experienced by the cardiac tissues.

6 Conclusion

Auxetic materials are materials with the unique property of having a negative Poisson's ratio. The selection of the appropriate base material and its structuring is essential to achieve the desired properties. Other properties of auxetic materials include high shear resistance, indentation resistance, fracture resistance, and energy absorption. Furthermore, their unique structures allow auxetic materials to attain synclastic behavior, variable permeability/porosity, and lower weight density. In most applications, the structuring of the material is the major factor in determining the properties and performance of the auxetic material. Thereby, the optimization of the geometric properties of the structure is essential to impart the desired properties. The types major of auxetic structures include re-entrant structures, rotating (semi-) rigid structures, crumpled sheets, perforated sheets, chiral structures, and auxetic gels. The selection of the structure type depends on the intended application of the auxetic material. The most commonly used configuration is the re-entrant type because of its stability, ease of fabrication, and good performance. The fabrication method depends upon the structure of the auxetic material, and the base material used. Additive manufacturing-based techniques such as Stereolithography (SLA), Digital Light Processing (DLP), Fused Deposition Modeling (FDM), Melt Electrowriting (MEW), and Multiphoton Lithography (MPL) are being used by research groups to develop the prototypes of auxetic materials. The advancement in fabrication techniques is essential for the development and commercialization of auxetic materials. Auxetic materials have enormous potential in the biomedical field because they can be imparted properties that very closely resemble human tissues and other body parts. Therefore, they are being explored to be used as implants, stents, and cardiac patches in the human body. Auxetic materials are rapidly being used as scaffolds in tissue engineering due to their porous structure and controllable Young's modulus and Poisson's ratio.

References

1. Veronda DR, Westmann RA (1970) Mechanical characterization of skin—finite deformations. J Biomech 3:111
2. Williams JL, Lewis JL (1982) Properties and an anisotropic model of cancellous bone from the Proximal Tibial Epiphysis. J Biomech Eng 104:50
3. Derrouiche A, Zaïri F, Zaïri F (2019) A chemo-mechanical model for osmo-inelastic effects in the annulus fibrosus. Biomech Model Mechanobiol 18:1773
4. Timmins LH, Wu Q, Yeh AT et al (2010) Structural inhomogeneity and fiber orientation in the inner arterial media. Am J Physiol Circ Physiol 298:H1537
5. Kim Y, Son KH, Lee JW (2021) Auxetic structures for tissue engineering scaffolds and biomedical devices. Materials (Basel) 14:6821
6. Gatt R, Vella Wood M, Gatt A et al (2015) Negative Poisson's ratios in tendons: an unexpected mechanical response. Acta Biomater 24:201

7. Ghasemi-Mobarakeh L, Prabhakaran MP, Morshed M et al (2008) Electrospun poly(ε-caprolactone)/gelatin nanofibrous scaffolds for nerve tissue engineering. Biomaterials 29:4532
8. Lakes RS, Witt R (2002) Making and characterizing negative Poisson's Ratio materials. Int J Mech Eng Educ 30:50
9. Wallbanks M, Khan MF, Bodaghi M et al (2022) On the design workflow of auxetic metamaterials for structural applications. Smart Mater Struct 31:023002
10. Carneiro VH, Meireles J, Puga H (2013) Auxetic materials—a review. Mater Sci 31:561
11. Yang W, Li Z-M, Shi W et al (2004) Review on auxetic materials. J Mater Sci 39:3269
12. Critchley R, Corni I, Wharton JA et al (2013) A review of the manufacture, mechanical properties and potential applications of auxetic foams. Phys Status Solidi 250:1963
13. Li Z, Wang KF, Wang BL (2021) Indentation resistance of brittle auxetic structures: combining discrete representation and continuum model. Eng Fract Mech 252:107824
14. Tretiakov KV, Wojciechowski KW (2012) Elasticity of two-dimensional crystals of polydisperse hard disks near close packing: surprising behavior of the Poisson's ratio. J Chem Phys 136:204506
15. Bezazi A, Boukharouba W, Scarpa F (2009) Mechanical properties of auxetic carbon/epoxy composites: static and cyclic fatigue behaviour. Phys Status Solidi 246:2102
16. Chen Y, Qian F, Zuo L et al (2017) Broadband and multiband vibration mitigation in lattice metamaterials with sinusoidally-shaped ligaments. Extrem Mech Lett 17:24
17. Yang S, Chalivendra VB, Kim YK (2017) Fracture and impact characterization of novel auxetic Kevlar®/Epoxy laminated composites. Compos Struct 168:120
18. Lakes R (1987) Foam structures with a negative Poisson's Ratio. Science 235(80):1038
19. Alderson A, Alderson KL, Chirima G et al (2010) The in-plane linear elastic constants and out-of-plane bending of 3-coordinated ligament and cylinder-ligament honeycombs. Compos Sci Technol 70:1034
20. Wang Z, Luan C, Liao G et al (2020) Progress in Auxetic mechanical metamaterials: structures, characteristics, manufacturing methods, and applications. Adv Eng Mater 22:2000312
21. Attard D, Casha A, Grima J (2018) Filtration properties of Auxetics with rotating rigid units. Materials (Basel) 11:725
22. Evans KE (2001) Auxetic polymers. Membr Technol 2001:9
23. Love AEH (1944) A treatise on the mathematical theory of elasticity. Dover, New York
24. Keskar NR, Chelikowsky JR (1992) Negative Poisson ratios in crystalline SiO_2 from first-principles calculations. Nature 358:222
25. Grima JN, Jackson R, Alderson A, Evans KE (2000) Do Zeolites Have Negative Poisson's Ratios? Adv Mater 12:1912
26. Williams JJ, Smith CW, Evans KE et al (2007) Off-Axis Elastic Properties and the effect of Extraframework species on structural flexibility of the NAT-Type Zeolites: simulations of structure and elastic properties. Chem Mater 19:2423
27. Sanchez-Valle C, Lethbridge ZAD, Sinogeikin SV et al (2008) Negative Poisson's ratios in siliceous zeolite MFI-silicalite. J Chem Phys 128:184503
28. Wang H, Li X, Li P, Yang J (2017) δ-Phosphorene: a two dimensional material with a highly negative Poisson's ratio. Nanoscale 9:850
29. Mortazavi B, Shahrokhi M, Makaremi M, Rabczuk T (2017) Anisotropic mechanical and optical response and negative Poisson's ratio in Mo 2 C nanomembranes revealed by first-principles simulations. Nanotechnology 28:115705
30. Deng B, Hou J, Zhu H et al (2017) The normal-auxeticity mechanical phase transition in graphene. 2D Mater 4:021020
31. Zhou L, Zhuo Z, Kou L et al (2017) Computational dissection of two-dimensional rectangular titanium Mononitride TiN: Auxetics and promises for Photocatalysis. Nano Lett 17:4466
32. Baughman RH (2003) Auxetic materials: avoiding the shrink. Nature 425:667
33. Wang N (2014) Auxetic nuclei. Nat Mater 13:540
34. Kolken HMA, Zadpoor AA (2017) Auxetic mechanical metamaterials. RSC Adv 7:5111

35. Masters IG, Evans KE (1996) Models for the elastic deformation of honeycombs. Compos Struct 35:403
36. Grima JN, Evans KE (2006) Auxetic behavior from rotating triangles. J Mater Sci 41:3193
37. Grima JN, Evans KE (2000) Auxetic behavior from rotating squares. J Mater Sci Lett 19:1563
38. Chan N, Evans KE (1999) The mechanical properties of conventional and Auxetic foams. Part II: Shear. J Cell Plast 35:166
39. Choi JB, Lakes RS (1992) Non-linear properties of polymer cellular materials with a negative Poisson's ratio. J Mater Sci 27:4678
40. Chan N, Evans KE (1999) The mechanical properties of conventional and Auxetic foams. Part I: compression and tension. J Cell Plast 35:130
41. Gibson LJ, Ashby MF (1997) Cellular solids, structure and properties. Cambridge University Press
42. Choi JB, Lakes RS (1995) Nonlinear analysis of the Poisson's Ratio of negative Poisson's Ratio foams. J Compos Mater 29:113
43. Grima J N, Alderson A, Evans KE (2005) An alternative explanation for the negative Poisson's Ratios in Auxetic foams. J Phys Soc Japan 74:1341
44. Chan N, Evans KE (1997) Fabrication methods for auxetic foams. J Mater Sci 32:5945
45. Scarpa F, Pastorino P, Garelli A et al (2005) Auxetic compliant flexible PU foams: static and dynamic properties. Phys Status Solidi 242:681
46. Quadrini F, Bellisario D, Campoli L et al (2015) Auxetic epoxy foams produced by solid state foaming. J Cell Plast 52
47. Smith FC, Scarpa FL, Burriesci G (2002) Simultaneous optimization of the electromagnetic and mechanical properties of honeycomb materials. In: Davis LP (ed), pp 582–591
48. Scarpa F, Tomlin PJ (2000) On the transverse shear modulus of negative Poisson's ratio honeycomb structures. Fatigue Fract Eng Mater Struct 23:717
49. Whitty J, Nazare F, Alderson A (2002) Modelling the effects of density variations on the in-plane Poisson's ratios and Young's Moduli of periodic conventional and re-entrant honeycombs—Part 1: Rib thickness variations. IMRI J Artic, 21
50. Yang DU, Lee S, Huang FY (2003) Geometric effects on micropolar elastic honeycomb structure with negative Poisson's ratio using the finite element method. Finite Elem Anal Des 39:187
51. Bezazi A, Scarpa F, Remillat C (2005) A novel centresymmetric honeycomb composite structure. Compos Struct 71:356
52. Wan H, Ohtaki H, Kotosaka S, Hu G (2004) A study of negative Poisson's ratios in auxetic honeycombs based on a large deflection model. Eur J Mech A/Solids 23:95
53. Larsen UD, Signund O, Bouwsta S (1997) Design and fabrication of compliant micromechanisms and structures with negative Poisson's ratio. J Microelectromechanical Syst 6:99
54. Gaspar N, Ren X, Smith C et al (2005) Novel honeycombs with auxetic behaviour. Acta Mater 53:2439
55. Grima JN, Gatt R, Alderson A, Evans KE (2005) On the potential of connected stars as auxetic systems. Mol Simul 31:925
56. Álvarez Elipe JC, Díaz Lantada A (2012) Comparative study of auxetic geometries by means of computer-aided design and engineering. Smart Mater Struct 21:105004
57. Evans KE, Nkansah MA, Hutchinson IJ (1994) Auxetic foams: modelling negative Poisson's ratios. Acta Metall Mater 42:1289
58. Wang Y, Wang L, Ma Z, Wang T (2016) Parametric analysis of a cylindrical negative Poisson's ratio structure. Smart Mater Struct 25:035038
59. Li D, Dong L, Lakes RS (2016) A unit cell structure with tunable Poisson's ratio from positive to negative. Mater Lett 164:456
60. Yang L, Harrysson O, West H, Cormier D (2012) Compressive properties of Ti–6Al–4V auxetic mesh structures made by electron beam melting. Acta Mater 60:3370
61. Bückmann T, Stenger N, Kadic M et al (2012) Tailored 3D mechanical metamaterials made by dip-in direct-laser-writing optical lithography. Adv Mater 24:2710

62. Critchley R, Corni I, Wharton JA, et al (2013) The Preparation of Auxetic foams by three-dimensional printing and their characteristics. Adv Eng Mater n/a
63. Wang K, Chang Y-H, Chen Y et al (2015) Designable dual-material auxetic metamaterials using three-dimensional printing. Mater Des 67:159
64. Shen J, Zhou S, Huang X, Xie YM (2014) Simple cubic three-dimensional auxetic metamaterials. Phys Status Solidi 251:1515
65. Yang L, Harrysson O, West H, Cormier D (2015) Mechanical properties of 3D re-entrant honeycomb auxetic structures realized via additive manufacturing. Int J Solids Struct 69–70:475
66. Wang X-T, Wang B, Li X-W, Ma L (2017) Mechanical properties of 3D re-entrant auxetic cellular structures. Int J Mech Sci 131–132:396
67. Caddock BD, Evans KE (1989) Microporous materials with negative Poisson's ratios. I. Microstructure and mechanical properties. J Phys D Appl Phys 22:1877
68. Evans KE, Caddock BD (1989) Microporous materials with negative Poisson's ratios. II. Mechanisms and interpretation. J Phys D Appl Phys 22:1883
69. Alderson A, Evans KE (1995) Microstructural modelling of auxetic microporous polymers. J Mater Sci 30:3319
70. Pickles AP, Alderson KL, Evans KE (1996) The effects of powder morphology on the processing of auxetic polypropylene (PP of negative poisson's ratio). Polym Eng Sci 36:636
71. Alderson KL, Fitzgerald A, Evans KE (2000) The strain dependent indentation resilience of auxetic microporous polyethylene. J Mater Sci 35:4039
72. Grima JN, Alderson A, Evans KE (2004) Negative Poisson's Ratios from rotating rectangles. Comput Methods Sci Technol 10:137
73. Grima JN, Zammit V, Gatt R et al (2007) Auxetic behaviour from rotating semi-rigid units. Phys Status Solidi 244:866
74. Shan S, Kang SH, Zhao Z et al (2015) Design of planar isotropic negative Poisson's ratio structures. Extrem Mech Lett 4:96
75. Attard D, Grima JN (2012) A three-dimensional rotating rigid units network exhibiting negative Poisson's ratios. Phys Status Solidi 249:1330
76. Attard D, Grima JN (2008) Auxetic behaviour from rotating rhombi. Phys Status Solidi 245:2395
77. Attard D, Manicaro E, Grima JN (2009) On rotating rigid parallelograms and their potential for exhibiting auxetic behaviour. Phys Status Solidi 246:2033
78. Alderson K, Alderson A, Ravirala N et al (2012) Manufacture and characterisation of thin flat and curved auxetic foam sheets. Phys Status Solidi 249:1315
79. Ma X, Zachariah MR, Zangmeister CD (2012) Crumpled Nanopaper from GRAPHENE OXIDe. Nano Lett 12:486
80. Hall LJ, Coluci VR, Galvão DS, et al (2008) Sign change of Poisson's Ratio for carbon nanotube sheets. Science (80)320:504
81. Grima JN, Grech MC, Grima-Cornish JN et al (2018) Giant Auxetic behaviour in engineered graphene. Ann Phys 530:1700330
82. Grima JN, Winczewski S, Mizzi L et al (2015) Tailoring graphene to achieve negative Poisson's Ratio properties. Adv Mater 27:1455
83. Eidini M (2016) Zigzag-base folded sheet cellular mechanical metamaterials. Extrem Mech Lett 6:96
84. Liu S, Lu G, Chen Y, Leong YW (2015) Deformation of the Miura-ori patterned sheet. Int J Mech Sci 99:130
85. Grima JN, Gatt R (2010) Perforated sheets exhibiting negative Poisson's Ratios. Adv Eng Mater 12:460
86. Gao D, Wang S, Zhang M, Zhang C (2021) Experimental and numerical investigation on in-plane impact behaviour of chiral auxetic structure. Compos Struct 267:113922
87. Mizzi L, Azzopardi KM, Attard D et al (2015) Auxetic metamaterials exhibiting giant negative Poisson's ratios. Phys Status Solidi Rapid Res Lett 9:425

88. Grima JN, Mizzi L, Azzopardi KM, Gatt R (2016) Auxetic perforated mechanical metamaterials with randomly oriented cuts. Adv Mater 28:385
89. Carta G, Brun M, Baldi A (2016) Design of a porous material with isotropic negative Poisson's ratio. Mech Mater 97:67
90. Lakes R (1991) Deformation mechanisms in negative Poisson's ratio materials: structural aspects. J Mater Sci 26:2287
91. Grima JN, Gatt R, Farrugia P-S (2008) On the properties of auxetic meta-tetrachiral structures. Phys status solidi 245:511
92. Prall D, Lakes RS (1997) Properties of a chiral honeycomb with a Poisson's Ratio of—1. Int J Mech Sci 39:305
93. Alderson A, Alderson KL, Attard D et al (2010) Elastic constants of 3-, 4- and 6-connected chiral and anti-chiral honeycombs subject to uniaxial in-plane loading. Compos Sci Technol 70:1042
94. Mousanezhad D, Haghpanah B, Ghosh R et al (2016) Elastic properties of chiral, anti-chiral, and hierarchical honeycombs: a simple energy-based approach. Theor Appl Mech Lett 6:81
95. Bhullar SK, Cerkez I, Sezer A, Jun MBG (2015) Swelling behavior of hydrogels within Auxetic polytetrafluoroethylene jacket. Polym Plast Technol Eng 54:1787
96. Ma Y, Zheng Y, Meng H et al (2013) Heterogeneous PVA hydrogels with micro-cells of both positive and negative Poisson's ratios. J Mech Behav Biomed Mater 23:22
97. Kunwar P, Jannini AVS, Xiong Z et al (2020) High-resolution 3D printing of stretchable hydrogel structures using optical projection lithography. ACS Appl Mater Interfaces 12:1640
98. Mizzi L, Salvati E, Spaggiari A et al (2020) Highly stretchable two-dimensional auxetic metamaterial sheets fabricated via direct-laser cutting. Int J Mech Sci 167:105242
99. Melchels FPW, Feijen J, Grijpma DW (2010) A review on stereolithography and its applications in biomedical engineering. Biomaterials 31:6121
100. Warner JJ, Gillies AR, Hwang HH et al (2017) 3D-printed biomaterials with regional auxetic properties. J Mech Behav Biomed Mater 76:145
101. Wagner M, Chen T, Shea K (2017) Large shape transforming 4D Auxetic structures. 3D Print Addit Manuf 4:133
102. Zheng X, Lee H, Weisgraber TH, et al (2014) Ultralight, ultrastiff mechanical metamaterials. Science (80)344:1373
103. Revilla-León M, Özcan M (2019) Additive manufacturing technologies used for processing polymers: current status and potential application in prosthetic dentistry. J Prosthodont 28:146
104. Wu J, Zhao Z, Kuang X et al (2018) Reversible shape change structures by grayscale pattern 4D printing. Multifunct Mater 1:015002
105. Li S, Hu M, Xiao L, Song W (2020) Compressive properties and collapse behavior of additively-manufactured layered-hybrid lattice structures under static and dynamic loadings. Thin-Walled Struct 157:107153
106. Yang C, Vora HD, Chang Y (2018) Behavior of auxetic structures under compression and impact forces. Smart Mater Struct 27:025012
107. Mazzoli A (2013) Selective laser sintering in biomedical engineering. Med Biol Eng Comput 51:245
108. Kapnisi M, Mansfield C, Marijon C et al (2018) Auxetic cardiac patches with tunable mechanical and conductive properties toward treating myocardial infarction. Adv Funct Mater 28:1800618
109. Geng LC, Ruan XL, Wu WW et al (2019) Mechanical properties of selective laser sintering (SLS) additive manufactured chiral Auxetic cylindrical stent. Exp Mech 59:913
110. Meng Z, He J, Cai Z et al (2020) Design and additive manufacturing of flexible polycaprolactone scaffolds with highly-tunable mechanical properties for soft tissue engineering. Mater Des 189:108508
111. Paxton NC, Lanaro M, Bo A et al (2020) Design tools for patient specific and highly controlled melt electrowritten scaffolds. J Mech Behav Biomed Mater 105:103695
112. Olvera D, Sohrabi Molina M, Hendy G, Monaghan MG (2020) Electroconductive Melt Electrowritten patches matching the mechanical anisotropy of human Myocardium. Adv Funct Mater 30:1909880

113. Flamourakis G, Spanos I, Vangelatos Z et al (2020) Laser-made 3D Auxetic metamaterial scaffolds for tissue engineering applications. Macromol Mater Eng 305:2000238
114. Grima JN, Attard D, Gatt R, Cassar RN (2009) A Novel Process for the manufacture of Auxetic foams and for their re-conversion to conventional form. Adv Eng Mater 11:533
115. Kim HW, Kim TY, Park HK et al (2018) Hygroscopic Auxetic on-skin sensors for easy-to-handle repeated daily use. ACS Appl Mater Interfaces 10:40141
116. Han S, Jung S, Jeong S et al (2020) High-performance, biaxially stretchable conductor based on Ag composites and hierarchical auxetic structure. J Mater Chem C 8:1556
117. Martz EO, Lakes RS, Goel VK, Park JB (2005) Design of an artificial intervertebral disc exhibiting a Negative Poisson's Ratio. Cell Polym 24:127
118. Jiang Y, Shi K, Zhou L et al (2023) 3D-printed auxetic-structured intervertebral disc implant for potential treatment of lumbar herniated disc. Bioact Mater 20:528
119. Yao Y, Yuan H, Huang H et al (2021) Biomechanical design and analysis of auxetic pedicle screw to resist loosening. Comput Biol Med 133:104386
120. Ghavidelnia N, Bodaghi M, Hedayati R (2020) Femur Auxetic meta-Implants with Tuned Micromotion distribution. Materials (Basel) 14:114
121. Kolken HMA, Janbaz S, Leeflang SMA et al (2018) Rationally designed meta-implants: a combination of auxetic and conventional meta-biomaterials. Mater Horizons 5:28
122. Verdonschot N, Huiskes R (1996) Mechanical effects of stem cement interface characteristics in total hip replacement. Clin Orthop Relat Res 329:326
123. Chamay A, Tschantz P (1972) Mechanical influences in bone remodeling. Experimental research on Wolff's law. J Biomech 5:173
124. Sousa JE, Serruys PW, Costa MA (2003) New frontiers in cardiology. Circulation 107:2274
125. Ali MN, Busfield JJC, Rehman IU (2014) Auxetic oesophageal stents: structure and mechanical properties. J Mater Sci Mater Med 25:527
126. Bhullar SK, Ko J, Cho Y, Jun MBG (2015) Fabrication and Characterization of nonwoven Auxetic polymer stent. Polym Plast Technol Eng 54:1553
127. Lee JW, Soman P, Park JH et al (2016) A tubular biomaterial construct exhibiting a negative Poisson's Ratio. PLoS ONE 11:e0155681
128. Hamzehei R, Rezaei S, Kadkhodapour J et al (2020) 2D triangular anti-trichiral structures and auxetic stents with symmetric shrinkage behavior and high energy absorption. Mech Mater 142:103291
129. Lin C, Zhang L, Liu Y et al (2020) 4D printing of personalized shape memory polymer vascular stents with negative Poisson's ratio structure: a preliminary study. Sci China Technol Sci 63:578
130. Chen Y-W, Wang K, Ho C-C et al (2020) Cyclic tensile stimulation enrichment of Schwann cell-laden auxetic hydrogel scaffolds towards peripheral nerve tissue engineering. Mater Des 195:108982
131. Soman P, Lee JW, Phadke A et al (2012) Spatial tuning of negative and positive Poisson's ratio in a multi-layer scaffold. Acta Biomater 8:2587

Chapter 2
Advances in Orthotic Prosthetic Design: Challenges and Applications

Arnab Chanda, Biswarup Mukherjee, and Subhodip Chatterjee

1 Introduction

Devices like prostheses and orthoses assist disabled individuals and reach out to their biomechanical requirements. Health clinicians utilize prosthetics to restore lost lower limb or upper-limb body parts. As an illustration, consider the artificial limb socket, a cup-shaped device that adjusts around an amputee's damaged limb and moving mechanical loads from the body appendages to the prosthesis. Braces and other similar terms for orthoses hold up and modify the human musculoskeletal system's structure and functional characteristics. Depending on the part of the body that is afflicted, orthoses can be categorized as top half, spinal, or lower-limb orthoses. They can also be termed after the joints they support, such as ankle–foot orthoses, lumbar orthoses, and wrist orthoses [1]. The fact that both orthotic (orthosis) and prosthetic (prosthesis) devices are made to support a particular body component to help the patient walk properly is one of their similarities. Both orthotics and prosthetics should meet the same ideal criteria, including being lightweight, rigid enough to resist shear stress, strong enough to withstand impact, affordable, and simple to connect to and remove from the injured body part. The primary distinction betwixt an orthotic device and a prosthetic device is that the former is applied to entirely substitute the missing limb, forasmuch as the latter is used to improve the functionality of limbs with defects or abnormalities that prevent proper function. Pre-engineered orthotics and prosthetics are more commonly accessible and less expensive than custom products on the market, but customized products that take the most important

A. Chanda (✉) · B. Mukherjee · S. Chatterjee
Centre for Biomedical Engineering, Indian Institute of Technology Delhi, New Delhi 110016, India
e-mail: arnab.chanda@cbme.iitd.ac.in

A. Chanda · B. Mukherjee
Department of Biomedical Engineering, All Indian Institute of Medical Sciences Delhi, New Delhi 110029, India

A. Chanda et al. (eds.), *Materials for Biomedical Simulation*, Materials Horizons: From Nature to Nanomaterials, https://doi.org/10.1007/978-981-99-5064-5_2

component in user satisfaction is how well a product fits a patient's body given their individual qualities [2]. Plaster casting which is a highly personalized, patient-centered procedure, is the most common and conventional method of producing custom orthoses and prostheses.

When compared to traditional manufacturing, additive manufacturing drastically lowers material waste, accelerates the fabrication process, and does away with the majority of human tasks that require skill. A patient in need of a prosthetic goes to the orthopedic or trauma specialist to have the necessary human body measurements taken in the traditional fabrication process. Plaster bandages are applied to the body part that is injured in order to fabricate a cast mold. The negative cast mold is subsequently filled with plaster to create a positive mold. The next step is to perform the heating and vacuum-forming of thermoplastic sheets (often polypropylene or polyethylene) onto the positive plaster mold in order to produce the prosthesis or orthosis. After the sheets have cooled and been cut to the proper shape, they are then trimmed. The plaster cast may be altered, or another element may be added, depending on the loading on the human body's sensitive and bearing parts. Then after, straps and accessories are added to complete the fabrication. The patient must go to the fitting session. Most of the time, more alterations are needed to assure the product's comfort and functionality. This process wastes materials and costs a lot of money in labor and time.

The ability and experience of the prosthetist or orthotist have a significant impact on the quality of the products [3]. Complex structures may be produced with additive manufacturing while reducing money and labor costs. The adaptability of additive manufacturing enables customization for unique applications or taking into account individual traits. It opens up new possibilities for design flexibility, minimization of resource surplus and waste, and cost effectiveness in creating unique products. Precision replicas of current products are possible because of additive manufacturing [4] and enable weight reduction while increasing functional performance. Additionally, the incorporation of AM functions might lessen the requirement for assembly processes [5].

2 Historical Development of Orthotics and Prosthetics

Since the art of making splints and braces has been used on limbs with fractures. They were discovered in 20th-century excavations by the Hearst Egyptian Expedition of the University of California led by Dr. George A. Reisner, the history of orthotics can be tracked back to the ancient Egyptian age, which is thought to have occurred around 5000 years ago. The 5th Egyptian Dynasty, which corresponds to 2750–2625 B.C., was said to have produced these splints, making them the earliest splints ever discovered. By employing the radiologic study of two mummies, Brier and his team made another significant discovery in 2015 that further establishes the existence of prosthetics dating back to the ancient Egyptian era [6]. Following the discovery of two more artificial toes, which are currently on display at the British

Museum, the artificial toe's operation was further explained. Since 1981, both the toes have been housed in the museum. The Greco-Roman World was still prospering between the years 1000 BC and 476 AD, and throughout this time, medical practices had a considerable impact on modern medicine. During this time, many well-known doctors were born, including Hippocrates, Herophilos, Dioscorides, and Galen of Pergamon. The "Father of Medicine," Hippocrates, was the most notable physician who had a significant impact on the evolution of medical procedures. Hippocrates employed wooden and leather splints to treat a fractured tibia from about 460 BC until 370 BC [7]. Two leather rings were used to make the splints, one of which was wrapped around the knee and the other over the ankle. A long cherry wood rod served as the link connecting the two rings. In order to give the ankle unfettered movement and a little skin exposed to facilitate scrutiny when necessary, the rods were positioned lateral to the ankle and knee. The world began the Renaissance era following the mediaeval era, which signifies the change from the Middle Ages to Modernity. This period lasted from about the year 1500 to 1800. Ambroise Pare, a French surgeon, invented the modern amputation techniques in the middle to late 1500s. He then developed prosthetic devices with characteristics such an engineering component such as a flexible harness, knee clamp control, and others that are still present in modern devices today. His primary driving force behind developing prosthetics was to aid the soldiers who had experienced traumatic events during the war. Hugh Owen Thomas created the Thomas splint in 1876, which is used to treat lower limb abnormalities. The main goal of the straightforward design was to immobilize the lower body appendages. The splint was made of a leather strap attached to an inclined rod that extends from both sides of the waist to the bottom of the foot. Yates and Lehneis published the initial article on how to use thermoformed plastics to replace metal orthoses in the 1960s. Many researchers argued and continually contrasted conventional metal and thermoformed plastics at the beginning of its use as an orthotic material substitute. Nonetheless in the following research, it was discovered that plastic had more benefits than metal.

It is clean, light, comfortable, and quiet. Plastic orthoses are more aesthetically pleasing than metal orthoses since they can be worn below the user's clothing. Since then, thermoformed plastics have taken over as the primary material for orthotics and prosthetics. Nigel Ring conducted the first experimental tests on carbon fiber composites in 1966. It was well-known that composites are a strong-to-weight ratio material. In other words, depending on how it was created and developed, it has the durability of a metal and the lightness of plastic. As a result, aside from plastics, carbon composites have emerged as one of the most advantageous materials (Fig. 1).

3 User Needs Analysis and Current State-of-the-Art

A substantial revolution is currently taking place in the fields of rehabilitative and assistive robotics, where technologies are being developed to actively aid or restore legged movement to those with muscle impairments or weakness, neurologic injury,

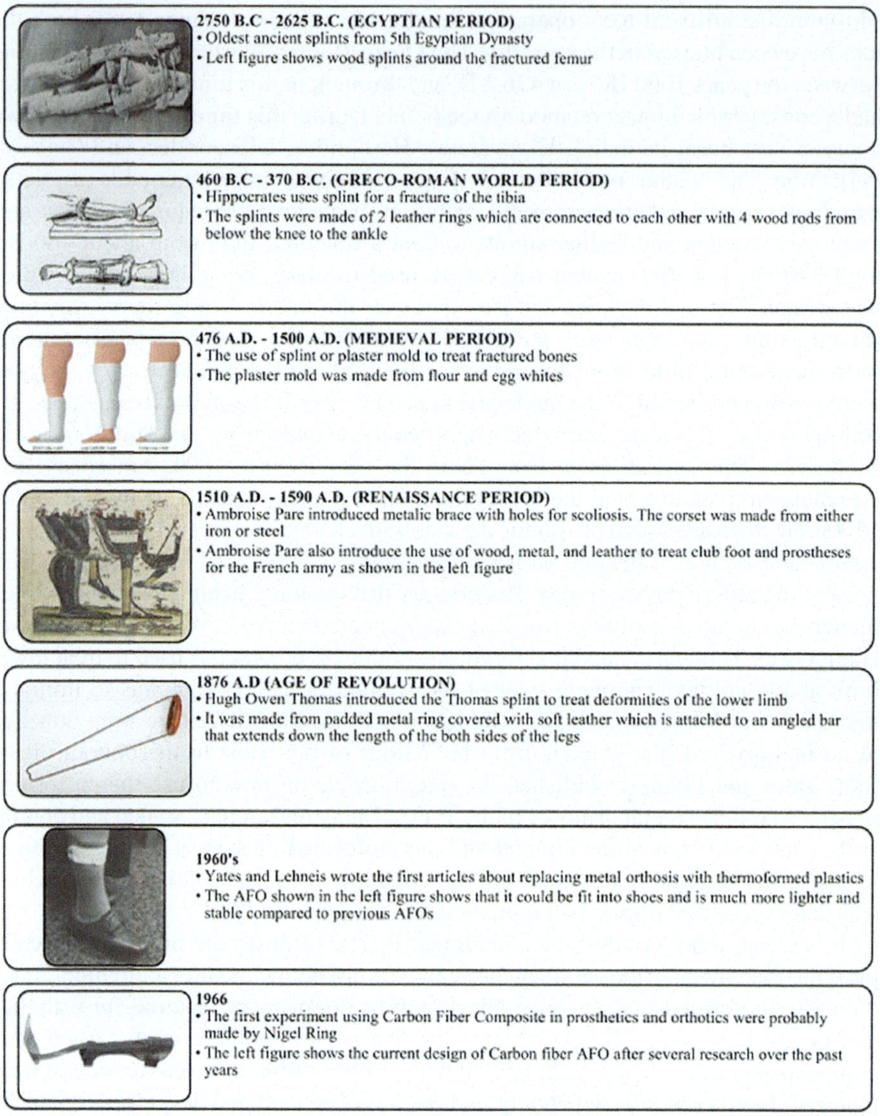

Fig. 1 Summary of the orthotic material development timeline [8]

or lower limb amputations. Energy-passive bionic devices have been used for a certain amount of time with varying levels of success [9]. For many disorders, passive devices provide a practical solution to enable efficient gait restoration, in part due to their relative simplicity, low initial cost, and strong construction. There are a number of inherent problems with these technologies, including their incapacity to generate mechanical power, their inability to automatically adapt to the user's changing needs, and their dearth of sensory feedback on the state of the limb and the device. For a perfect mental and physical connection between the user and the device, each of these elements must be present. Orthotics and prosthetics that are portable and intelligent have the potential to significantly increase a person's mobility and, consequently, their quality of life. The end-users will once again be able to engage in daily activities As these devices begin to approach the power output, efficiency, and adaptability of the limbs that they help or replace, they should be utilized for activities that demand total combined energetic output (e.g., stair ascending, sprinting, hopping) in the same ways as their able-bodied counterparts. In comparison to their passive counterparts, active prosthetics and orthotics may also be able to lower metabolic cost while increasing the self-selected gait speed [10–12]. These devices may help improve gait symmetry and lessen the risk of compensatory movements damaging the user's unaffected joints. Numerous mobile robotic systems for assisting and restoring human movement have been created. With improvements in computer technology, miniaturized sensors, energy storage, automatic pattern recognition and actuation it is possible that within the next ten years, numerous further active lower extremity prosthetics, robotic arms, and orthotic devices will be developed and made commercially available (Fig. 2).

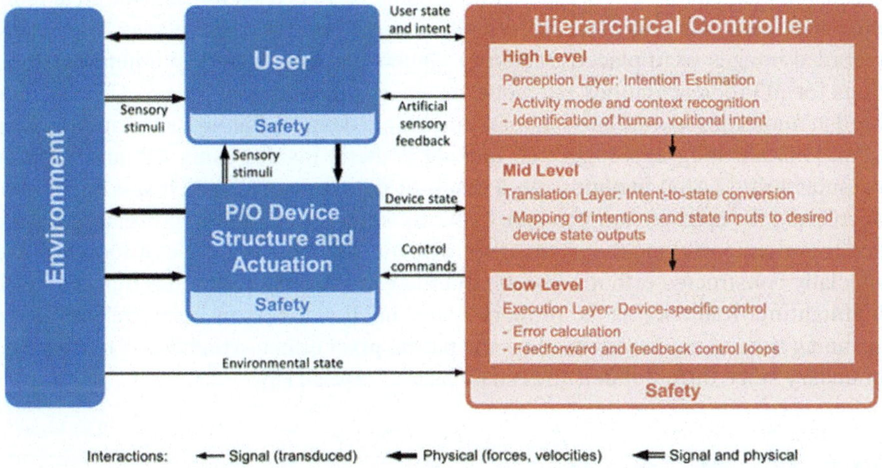

Fig. 2 Generalized control framework for active lower limb prosthesis and orthotics [13], CC-BY-4.0

4 Process of Development of Prosthesis and Orthotics for Patients

4.1 3D Anatomical Data Acquisition Technologies

New design guidelines for orthotic devices can be created by combining the use of rapid prototyping techniques with various methods for assessing and modeling the human body. The information can be represented as a data cloud, picture elements, or space coordinates of various physiological points, depending on the data gathering technique utilized. However, there are a number of acquisition ways that facilitate production employing fast prototyping techniques in the area of modeling of bionic devices. These acquisition methods include computer low dose CT scanning, polygonal scanning, and various visual aid movement capture systems.

4.1.1 Computed Tomography

The sophisticated tool of computed tomography (CT) is used for both surgical planning and diagnostic purposes. Historically, the axial or transverse planes were where recorded images were placed. Modern scanners can now create volumetric restorations for 3D representations by capturing images along many planes. CT has been used in numerous research to manufacture bionic devices. Taking for example, Tang et al. [14] not long ago suggested making diabetic insoles using CT and additive manufacturing (AM) methods. As a result of their research, which was capable of decreasing the apex plantar pressure by 33.67%, they were able to correlate pressure and tissue tension along the plantar foot with the treatment response of footwear and specially constructed orthotic inserts. But there are several difficulties that are worth highlighting. Radiation is the primary issue, and the dose is inversely related to the scanning time. Another downside is the partial pixel effect, which leads to a blurred boundary since different densities share shared pixels [15].

4.1.2 Three-Dimensional Scanning

3D scanning has emerged as the most convenient and comfortable method for capturing human morphology or the external contour. In order to establish the spatial location in three dimensions of all the points that together make up a material's surface, volumetric scanning systems use optical-based approaches. A CAD model is then obtained after using computer tools to recreate features from the point cloud. Currently, single image restoration, structured light technologies, lasers, and other stereo reconstruction methods are all used in 3D scanners for human assessment. Structured light and photonic technology are the most often utilized tools for reshaping the human body [16]. The laser method creates a laser dot or line with a portable device. A sensor-often a spot or charge-coupled device-measures the separation from the surface. Optical imaging techniques broadcast predetermined wave patterns onto the object moving using a projector-camera system. In the groundbreaking work of Chee Kai et al. [17], a volumetric method was chosen rather than employing more traditional methods like blockwork impressions, Magnetic resonance, and Computerized tomography for prosthetic modeling. Mavroidis et al. [18] developed patient-specific foot orthoses using 3D laser scanning. Through the use of designing software and a rapid prototyping tool, exterior characteristics of the patient's anatomy were optimized. Comparing the prototype to ankle–foot orthoses available for sale, the prototype more accurately fits the subject's anatomy. A drawback of this technology is its inability to record certain topographical human biological features with intricate wrinkles and folds, including the space between both the fingers when the palm is neutral, the back of the leg when flexed, or the armpits.

4.2 Rapid Prototyping Technologies for Orthotic Devices

In the field of orthotic devices, a well-known approach that is gaining greater attention is the substitution of traditional craft methods with Design software and computer-aided fabrication.

The following processes are required for customized production using rapid prototyping, according to Ciobanu et al. [19]: CAD modeling, translation to stereolithography format (STL), 3D scanning of the biological surface, 3 dimensional restoration, and, lastly, machining utilizing a specialized additive manufacturing machine (i.e., a 3D printer) operated by a system. For the construction of custom-fit orthotic devices, rapid prototyping offers benefits such as more design freedom, the production of functional requirements, higher accuracy and cost effectiveness, faster delivery, and enhanced user experience. A physical product is created layer by layer from a realistic virtual 3D CAD model using a rapid prototyping manufacturing technique [20]. An optical model of the component is created using a designing software and transformed during the rapid prototyping process into the STL file format, which is the default standard file format for RP systems. Depending on the type of the fabrication method, such as laser, printer, and extrusion technologies, additive manufacturing can be classed in a variety of ways [21]. There are numerous varieties of additive manufacturing techniques. In the early nineteenth century, Kruth [22] recommended using various additive manufacturing techniques, categorized based on the element employed for the model (Table 1). However, Paterson et al. [23] showed that only a few of them could be utilized in the creation of orthotic and prosthetic devices. Orthoses and prostheses are made utilizing solid-based processes like laminated object manufacturing (LOM) or liquid-based processes like stereolitography (SLA), solid ground curing (SGC), UV light curing (ULC), and ballistic particle manufacturing (BPM). However, 3D printing with a powder bed and inkjet head, selective laser sintering (SLS), and fused deposition modeling (FDM) are the most often employed

Table 1 Rapid Prototyping techniques available for prosthetics

Material	Process
Liquid base	• Stereolithography (SLA) • Solid ground curing (SGC) • UV light-curing (ULC) • Ballistic particle manufacturing (BPM)
Solid base	• Laminated object manufacturing (LOM) • Fused deposition modeling (FDM)
Powder base	• Selective laser sintering (SLS) • 3D printing (Polymer injection)

production techniques for orthotic and prosthetic devices (3DP). These methods show an ideal trade-off between price, turnaround time, accuracy, and comfort.

4.2.1 Fused Deposition Modeling (FDM)

In the FDM procedure, a quasi-material is pushed through an extrusion head that moves in the X and Y axes (see Fig. 3a) to produce a two-dimensional layer of the intended object. The moveable material pushing head is made up of two extrusion nozzles: one to hold the support material and the other to place the building materials. Typically, the extruder head fills the delimited zone created by the preceding extrusion by adhering to a predetermined pattern after extruding the perimeter of each layer. The support platform descends after the layer is finished, and another layer is extruded. Layer after layer, the technique goes on until the item is finished. Polycarbonate (PC) and Acrylonitrile Butadiene Styrene (ABS), or a combination of the two, are the most often used materials for FDM. These materials resemble thermoplastic polymers for injection molding in their characteristics. It is also possible to use other materials, including polymers or nylon-based compounds. The utilization of inexpensive materials is the main benefit of FDM technology. Tan et al. [24] were innovators in the application of utilized FDM for the manufacture of tibial prostheses and came to the conclusion that the prosthetics' functional qualities were appropriate for usage in clinical settings. On the other hand, production times are lengthy. Since then, more and more biomedical uses for FDM have emerged, including medication delivery systems, hand and facial prostheses, upper and lower limb orthoses, and hand prostheses.

4.2.2 Selective Laser Sintering (SLS)

In the 1990s, the DTM Corporation, which is now a part of 3D Systems, unveiled the first SLS technology. By employing a CO_2 laser to selectively fused granular polymer-based materials, such as nylon/polyamide, the SLS process produces volumetric solid objects or parts (Fig. 3b). In order to create a 2D profile, a CO_2 laser

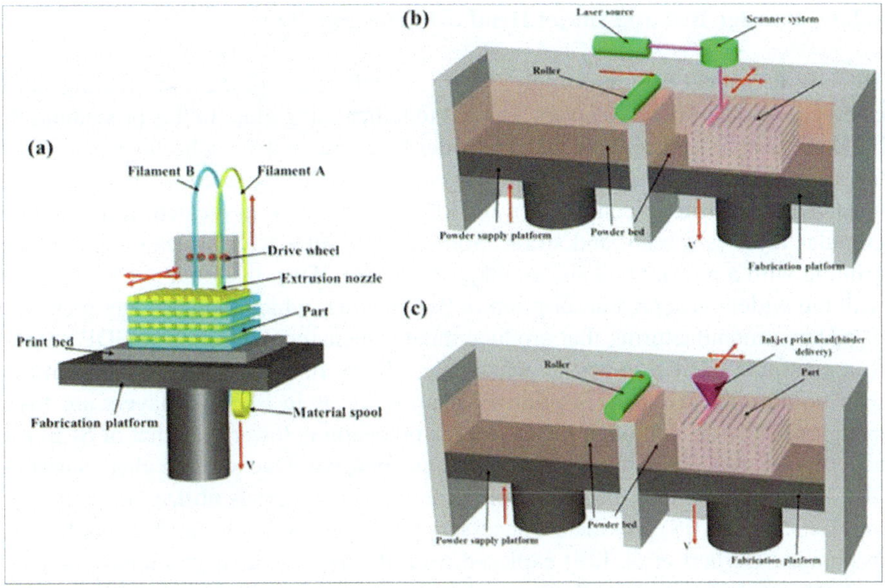

Fig. 3 Comparison among the rapid prototyping operations of fabricating prosthetics and orthosis **a** Fused deposition modeling (FDM), **b** Selective laser sintering (SLS), **c** 3DP [31], CC-BY-4.0

moves throughout the powder bed in the X and Y axes, selectively sintering designated portions. The platform descends, a fresh layer of powder is applied, and the sintering procedure is repeated when the 2D profile has already been finished. The procedure is afterwards referred to as a granular-based fusion technique. Generally speaking, all materials are thermoplastics, with polyamide 12 (PA), acrylonitrile butadiene styrene (ABS), and polycarbonate being the most popular ones (PC). The useability of the rehabilitation devices is improved by these materials' significant weight reduction. Schrank and Stanhope [25] assessed the accuracy of the SLS manufacturing method of foot prosthetics as an illustration of the use of this method in the fabrication of custom prosthetics. In this study, the divergence between the final product created using SLS and the CAD model was recorded using the Faroarm 3D scanner (accuracy of 25 m). The results revealed values that were less than 1.5 mm. Deckers et al. [26] created and evaluated an SLS-based AFO, underlining the necessity of accurately characterising the AFO's mechanical attributes such as strength, fatigue, and impact resistance. Following testing of a polyamide-based orthosis produced using SLS, Vasiliauskaite et al. [27] came to the conclusion that the features were comparable to those of a polypropylene orthosis formed by the process of thermoforming, with the first being more rigid than the second but still suitable for reclamation.

4.2.3 Powder Bed and Inkjet Head 3D Printing: 3DP

Three-dimensional printing, often known as 3DP, is the process of creating fabricated goods out of powder layers adhered together using glue. In this procedure, the build platform is first covered with a powder layer. Second, by adhering to a textured layer in the horizontal plane, a liquid agent is selectively placed by an inkjet print head. The platform descends when the 2D pattern has been created, following the granular layer is spread, and so on. Some people refer to this technology as "3D printing with a powder bed and an inkjet head" (or 3DP). It should not be mistaken with the widely accepted description of 3D printing, which includes any processes of additive manufacturing that produce three-dimensional things. The 3DP process is comparable to SLS in certain ways (Fig. 3b, c). In 3DP, the material is infused with liquid adhesive using a printing head, whereas in SLS, the layers are fused using a CO_2 laser. Although this method's precision is lower than that of SLS, it is nevertheless popular since it is quick and inexpensive. Due to these characteristics, 3DP already dominates the prototyping sector. The materials utilized (mostly thermoplastics like ABS) provide the necessary characteristics to be used for application in bionics. Herbert et al. [28] explored even if this technique was appropriate for producing functioning prostheses, and they suggested that, despite the low fabrication levels, patients preferred 3DP machine-made prostheses (Corporation Z402) over conventionally created ones. Unfortunately, the resistance was not investigated in that investigation, so the product's durability is uncertain. Utilizing this technology, Saijo et al. [29] created patient-specific maxillofacial implants and reported a decrease in procedure times. Because of its digital accuracy, control, and adaptability, The use of 3DP in bioengineering and recuperative medicine is particularly exciting., according to Ventola [30].

5 Materials Used for Fabricating Prosthetics and Orthotics

Figure 4 illustrates how materials used for orthotics and prosthetics have changed over time, progressing from wood, metal, and leather to plastic and carbon fiber composite.

5.1 Wood Prosthetics and Orthotics

Wood orthoses have been around since the dawn of humanity. Wood has several advantages over other materials for making orthoses, including renewability, machinability, superior strength to weight ratio, resistance to rusting, and aesthetic appeal. The disadvantage is that wood's characteristics are extremely varied. The species of wood, the amount of moisture in it, or even how the load is oriented against the grain, can all have an impact on how variable the mechanical properties of

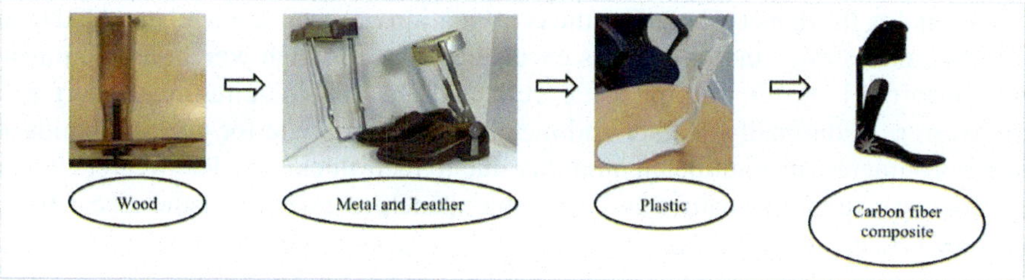

Fig. 4 Different types of materials used in orthotics and prosthetics [8]

wood are. Tension, compression, flexure, and finally shrinking and expanding make up the mechanical properties of wood. Typically, woods range in density from 160 kg/m^3 to 1350 kg/m^3. The bark of an oak tree, an oak tree's cork, or balsa wood were the most often utilized categories of wood. An orthosis must have adequate tensile strength, compressive strength, and flexural strength in order to assure welfare and stability. The two types of wood's tensile and compressive strengths are those that are perpendicular to the grain and those that are parallel to the grain, respectively. The strength is greatest when tension is exerted orthogonal to the grain. Consequently, depending on the wood species and its mechanical strength, the material may or may not be suitable for use as a prosthetics material. The type of wood that will be used will depend greatly on the prosthetics that will be created.

5.2 Metal and Leather Prosthetics and Orthotics

Leather is one of the earliest materials that has ever been utilized. Leather is made from animal skin, which is subsequently chemically synthesized during the tanning procedure. The tanning process will make the skin well built, more expandable, and more resilient. The cuffs that keep the appendages in place were created to mimic textiles or straps with laces, in contrast to the leathers used in the prosthetic design, which will always resemble a pair of shoes. High tensile strength, tear resistance, and efficient heat insulation are typical properties of modified leather. The drawbacks of utilizing leather are its limited biodegradability and the pollution that chemical wastes from each tanning operation will generate. The density of leather is 860 kg/m^3 on average, roughly. In order to endure the stress exerted towards the lower limbs, leather was typically combined with metal supports while making lower limb prosthetics. Metal alloys, steels, and any lightweight metal with a high tensile strength are the most often utilized metals in the manufacture of orthoses. Metals can exhibit densities ranging from 1700 kg/m^3 for a magnesium alloy to 20,000 kg/m^3 for a tungsten alloy. The fact that metal-leather orthoses frequently have a modular construction makes them advantageous to use. As a result, the shoe can be taken off of the metal supports or fixtures and be replaced with another shoe. This style of orthosis has the drawback of being bigger than any other orthosis and having no cosmetic appeal.

Additionally, the density including the combination of metal and leather added to an orthosis makes this kind of orthosis exceedingly having high weight and requiring more effort. As a result, walking while wearing these AFOs requires more effort from the wearer. Additionally, leather and metal are inappropriate for use as clothing in an environment with significant humidity due to their thickness. The wearers would experience extreme sweating and a terrible stench as a result of the sweat being absorbed by the leather.

5.3 Plastic Prosthetics and Orthotics

Most people are familiar with plastics as semi-synthetic materials. There are two main categories that it falls under: thermoplastics and thermosets. Thermoplastics are types of plastic that, when heated, can transform into a liquid and, when appropriately cooled, can solidify into a glassy substance. This substance is useful for making orthotics since it is simple to form into a plaster model which can be used to create prosthetics. Acrylic, Polypropylene (PP), Polyethylene (PE), and Nylon are a few thermoplastic examples. Thermosets are frequently in a liquid state before curing and can be molded into their final form once curing has taken place. Once thermoset has been molded, it cannot be changed. Thermosets including polyester resin, polyurethane, silicone, and epoxies are frequently utilized in orthotic devices. The fact that plastics are combustible, insignificant heat and electricity conductors, and do not corrode or rust make them advantageous for use in orthotics. These characteristics will guarantee that the manufactured orthosis won't hurt the users, whether physically or chemically. Because of its excellent moldability, the orthosis could be customized to just about any shape or size that would ideally fit the wearers. Plastics are also lightweight, strong, simple to color, and energy efficient. Thus, among orthoses made of various materials, those made of plastic are the lightest and most aesthetically beautiful. They also preserve their strength. Plastics have a density that ranges from 36 kg/m^3 to 2200 kg/m^3. Tensile strength ranges from 0.24 MPa to 170 MPa for this material. The modulus of elasticity of plastics can vary greatly depending on the kind of resins, reinforcing agents, and manufacturing methods employed. It is between 0.7 MPa and 4100 MPa. Plastic's indestructibility and potential for environmental pollution are its drawbacks. Despite the fact that the garbage is melted to eliminate it, the gas created during the melting process is still extremely detrimental to human health and may contribute to the weakening of the ozone layer. Plastics are typically lighter than the other materials used to make AFOs that are sold on the market. However, repeated use over a lengthy period of time also results in skin irritation. Plastic is currently the most desirable material due to the development of additive manufacturing. An orthosis is created via rapid manufacturing, which makes use of 3D printing and 3D CAD models. The 3D reconstruction created by CAD/CAM software will be used to print the orthosis layer by layer on the printer. Since a plaster cast model is not required as in the typical procedure, this could eventually lower the expense of producing the orthosis. The ability to redesign and do endless optimization also contributes to creating an advanced and more useful prosthetic while lowering the risk of a bad design.

5.4 *Carbon Fiber Reinforced Polymer (CFRP) Composite Prosthetics and Orthotics*

A composite is a substance that combines multiple materials with contrasting features in order to enhance each other and produce characteristics that are entirely separate from those of the constituent materials. The majority of orthotics products are constructed of polymer composites. The matrix of polymer composites is often made of thermoset or thermoplastic resins, with reinforcement elements including Aramid, carbon fiber, and glass fiber. Among all other polymer composites with the exception of that of plastics, CFRP is one of the most suitable materials utilized to make a prosthetic. Carbon fiber serves as the composite's reinforcement while polymers serve as the matrix. This kind of substance is as strong as metal while yet being lightweight. In comparison to other materials, it is also far more aesthetically pleasing. From 1500 kg/m^3 to 1600 kg/m^3, CFRPs have a density range. Tensile strength measurements range from 550 to 1100 MPa. Finally, it had an elastic modulus that ranged from 69,000 MPa to 150,000 MPa. The use of composites in this application has both benefits and drawbacks. The ability to reduce weight while maintaining good strength is one of the benefits. Additionally, it resists corrosion and wear. A composite mechanical behavior can be customized to meet the needs of a client or an application because of how it was manufactured. As a result, it can be used in applications where it is necessary to have two opposing features without surrendering any of them. The creation of composites is expensive, not always environmentally benign, and has a poor reusability rate as drawback (Table 2).

6 Finite-Element Modeling of Prosthetics and Orthotics

The performance of orthoses and prostheses can be predicted with great value using finite-element models. A paradigm that took the size and orientation of orthoses into account has been proposed by researchers. This substructure showed the dimensional precision of additive manufacturing but did not include biomechanical design optimization. In order to properly adjust and forecast the biomechanical properties of a passive-dynamic ankle–foot orthosis, a novel virtual functional prototyping procedure was created. It consists of finite-element model and digital model construction. The performance of the design was then evaluated after it was FDM-fabricated using medical-grade polycarbonate. The manufacturing precision, dimensional accuracy, and bending stiffness were all judged to be satisfactory. Instead of a structured procedure, this method was a collection of technologies. Any kind of surface form

Table 2 Summary of current prosthetics and orthosis materials

Materials	Advantages	Disadvantages
Wood	• It is renewable and having adequate strength to weight ratio • It does not undergo corrosion and is visually appealing	• Wood exhibits variable qualities
Metal and leather	• It is having high tensile strength and exhibits sufficient tear resistance • It is also having good heat insulation properties • It also comes in a customizable design	• Hazardous chemical wastes are created during the process of manufacturing • It has negligible biodegradability • It is overweight • It is having low visual value • Substantial energy is spent in fabricating them
Plastics	• It is simple to mold and speeds up manufacturing • Insignificant heat and electricity conduction ability • It is inflammable • It is having corrosion resistant properties • It is lightweight and strong	• It may result in excessive perspiration and foul odor • It is unbreakable and contributes to environmental degradation • The creation of hazardous gases during orthosis manufacturing is undesirable
Carbon fiber composite	• It exhibits good strength to weight ratio • It portrays substantial aesthetic value • It portrays fatigue resistance • It is corrosion resistant	• Involves high production cost and low rate of reusability • The design and manufacturing time is very extensive

imaging or scanning should work with the image processing and scanning package. For data with intricate inner structures, such as computerized tomography, electromagnetic resonance imaging, and acoustics, as well as point cloud surface data, such as photogrammetry, photogrammetry, and millimeter wave, geometric reconstruction would be acceptable. Based on the image processing and scanning data, the 3D geometry of the afflicted body part is recreated, and then a physical model of the prototype prosthetic is created. To create the initial design model, computerized manipulation based on the prosthetist's experience and fundamental design concepts is used. Models of the affected body part and the initial prosthetic design are further evolved into finite-element models and combined to simulate wearing and movement activities. For patient fitting and measurement trials, a realistic model of the prosthesis or orthosis' original design is used. Measurements of the biomechanical factors, such as the contact surface, contact pressure, shear force, temperature, and humidity, must be made during the experiments.

Motion analysis is done on people wearing orthoses or prostheses to assess kinetic and kinematic behavior in order for subsequent computational simulation and the creation of perimeter and load application conditions. The finite-element models are validated by comparing measurements taken at the contact interface with the outcomes of the computer analysis. In addition to the perimeter conditions, loading, and validation conditions that were obtained from the tests, the computer analysis also requires input from the biomechanical properties. Whenever the material for the design is confirmed, the product's material attributes are established. In the study of biomechanics, tissue attributes, particularly those of muscular tissues, present a problem. The mechanical characteristics of soft tissues can be measured in vivo using an ultrasound indentation device, which is simple and rapid to use. The inner workings of the body, contact behavior at the contact region, and biological details of a prosthesis or orthosis can all be learned through computational simulation. These numbers, along with the characteristics measured throughout the experiments, would be reviewed and compared in order to discover unjustifiable performance, such as stress distribution, excessive loading in a loading-sensitive spot, or limited deformation during motion. The model representing the basic design of the prosthesis or orthosis would undergo structural or material alteration if the finite-element analysis anticipated overall performance that was unrealistic. The next finite-element model would then be again meshed and assembled with the model of the affected human body part, and the same procedures would be repeated until the parameters indicate an acceptable and gratifying overall performance. Digital modification would be applied to the particular area, followed by the revised meshing, modified assembly, and movement simulation, if the results of the finite-element analysis only show specified illogical behavior, such as stress concentration in a local area. Finite-element simulation can react to such model changes quickly. The augmentation cycle would be continued until all investigated parameters of the computational prediction showed reasonability and were congruent with experimental data. Topology maximization would be used to redistribute materials such that the products would be lighter while still meeting the required strength (Fig. 5).

7 Neural Prosthetics and Signal Processing

By converting cortical cerebral activity into measurable signals, a new family of prosthetics aims to enable command by computers, prosthetic arms, and paralyzed upper-limbs. Only after the anticipated quality of life gain overcomes the possible dangers are neural prosthetic devices clinically viable. Internal body, electrode-based techniques have emerged as a key area of research because they offer much clearer signal quality and the promise for enhanced performance compared to exterior body options. Noninvasive techniques are appealing because of their lower surgical risk. For instance, the most advanced electrode-based system now used in laboratories can transport information at a rate of 6.5 b/s, which is far higher than previous invasive and noninvasive systems. However, intrusive approaches come at a high price and higher surgical risk. Congenital prosthetics based on sensors are therefore now a substantial strategy, with possible near-term applications being restricted to only the most seriously injured patients. It will be necessary to balance performance,

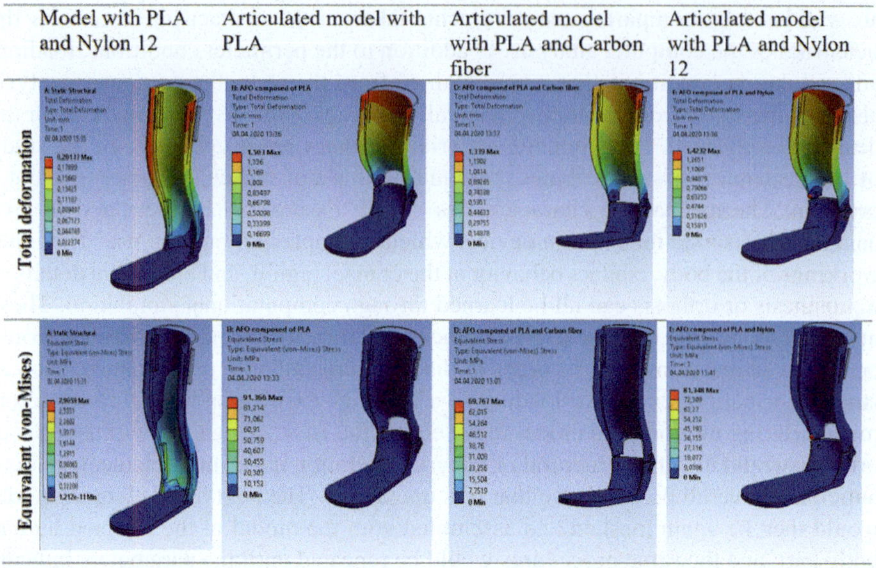

Fig. 5 Simulation results for the conventional and articulated AFO model [32],

risk, and cost by raising total bionic performance and lowering operative risk and device cost through system integration in order to go from research to mainstream usage [33, 34]. The performance of the prosthesis is enabled by high-quality brain signal monitoring and sophisticated signal processing techniques, including rigorous real-time action potential identification (peak sorting) and probabilistic movement decoding algorithms. These methods differ from others in that they can extract more distinct neurons with greater accuracy during the spike identification process, and they can incorporate more neural activity during the decoding phase while doing it more effectively. It is believed that with additional advancements in brain measuring and signal processing techniques, >10 b/s systems would be feasible. While not always the most accurate representation of a clinical setting, equipment-intensive, laboratory-based trials in which a controlled subject completes a strictly regulated task while being watched by a research scientist are a potent experimental platform. Clinical systems must be independent and able to recognize patient responses, specifically if neural activity genuinely reflects the desired movement, based solely on that neural activity. They cannot rely on trained operators or external control. Furthermore, rather than merely during the brief, distinct daily recording intervals utilized in present experimental techniques, prosthetic systems must offer these capacities reliably and constantly (24 h per day, every day). Unsupervized learning-based spike sorting algorithms eliminate the need for an expert operator and have the potential to produce robust, adaptable algorithms that can react on their own to changes in the brain recordings. Similar to this, decoding algorithms that automatically recognize neural states (whether movement is intended or not) do away with the requirement for outside cues to pinpoint times when significant neural activity is occurring.

By enabling system integration and getting rid of persistent transcutaneous linkages, prosthetic systems need to lower the risk of surgery and device costs. The ultimate goal is an implantable system with electrodes, wireless telemetry, digital post-processing, and cutting-edge functionality in a self-contained container with a significantly reduced size and no chronic tissue holes. However, in this method, a very constrained power budget must be followed for signal collecting and processing. A significant difficulty is the transmission of brain information away from the electrode implantation site. Current wireless networks can provide the necessary bandwidth, but their high power requirements make them impractical. It is imperative to reduce bandwidth in some way. There are several ways to accomplish this reduction, but many of them use lossy compression, which can compromise the performance of prosthetics. High performance signal processing methods can be employed to lower the necessary bandwidth by a factor of 106 in the implanted system while still meeting power requirements, with the aim of not sacrificing any prosthetic performance. Signal processing engineers face a difficult design problem due to the confluence of stringent power limits, aggressive performance goals, and demands for resilience and autonomy. New algorithms and implementations will be required in the future.

8 Chronic Electrode-Based Prostheses

Figure 6 depicts the fundamental design of motor and communicative prostheses (a). While communicative prostheses try to create an interconnecting route similar to "typing" on a computer, motor prostheses strive to restore neurological control to the paralyzed limb. A sample of these estimate (decoding) algorithms is illustrated in Fig. 6, and they leverage the link between such a motion and the neural response (tune) to produce the desirable output from just the brain activity. After that, the system can generate the requisite control signals to continuously move a paralyzed or bionic arm in space (arm prosthesis) or position a computer cursor over the necessary key on a computer (communication prosthesis). The plan activity, which is active from shortly after the reach objective is established until just before the movement starts, is specific to the movement's target. Movement activity, on the other hand, is present from just after the movement begins. Almost all of the neural activity detected in the motor and dorsal premotor (PMd) cortex is spiking activity. Although the local field potential has been demonstrated to be able to anticipate the direction of movement in other cortical regions, its function in M1 and PMd is still unknown. Communication prosthesis can be driven by decoding this information; they simply have to estimate the movement terminus. As the intended movement is to be replicated, the motor prosthesis must have movement activity. However, a planned activity can contribute to motor prosthesis by giving a priori information, such as a target estimation, that limits movement estimation.

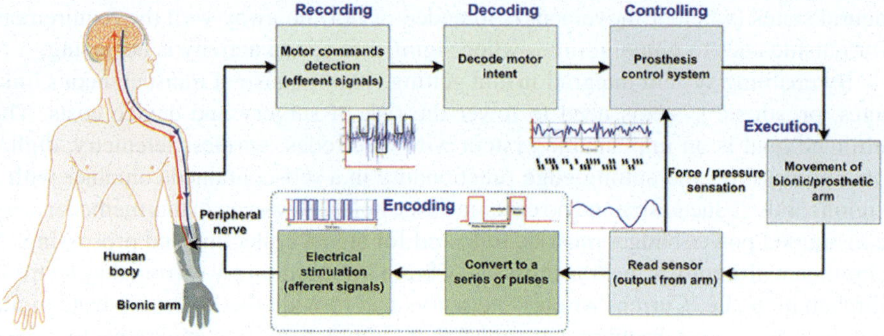

Fig. 6 Concept of bi-directional control for bionic arm systems [35], CC-BY-4.0

9 Application of Machine Learning in Prosthetics

For those who have had amputations, upper-body prosthetics are intended to replace bygone hand and related arm functions. The upper-body is complicated in structure and can make coordinated motions across numerous ways of freedom, making it difficult to recreate dexterous hand function. These ways include finger joint movements, forearm extension, and more. Robotic devices (equipped with actuators, circuits, and control systems) known as "myoelectric prostheses" are intended to imitate the movement and functionality of a biological arm and hand. A specially made socket is used to attach the bionic device to a user's remaining body appendages. Surface electrodes in the socket pick up EMG signals when the user actively contracts specific muscles in the residual limb. These muscle signals are then sent to the prosthetic controller, which filters the raw data, extracts signal features from them, and uses signal processing techniques and a control algorithm to identify the intended movement. The resulting control signals are then converted to electrical signals, and the device motors execute the commands. People with upper-limb amputations now have hopeful restorative movement alternatives thanks to years of developments in signal-actuated prosthetic control. However, there are several major limits to the accurate practical deciphering of motion intent from EMG signals, and it is still difficult to achieve accurate and natural prosthesis control. Customary myoelectric prosthesis control strategies can be broadly divided into two categories: 1) on/off, which allows for the binary closing and opening of a hand when EMG signals direct the triggering potential, or 2) quantifiable, which regulates the velocity of the opening and closing of a hand and enables much smoother movement control. With each of these standard strategies, the contraction of two distinct residual muscles normally initiates the "hand open" and "hand close" movements (with two different electrodes used to detect these opposing actions). Therefore, the activation of extra residual muscles is required to regulate additional joint movements. The amount of distinct muscle impulses in a user's remaining limb that can control each DOF limits the robustness of the on/off and proportional control techniques. Given that modern upper-part myoelectric prosthetics often offer higher DOF than the number of separate EMG signals that a device user may create, this constraint poses an operating problem. Additionally, to this restriction, the intrinsic unpredictability in EMG signals might result in inconsistent prosthesis control and unintentional movements of the prosthesis. Modifications in the user's limb posture, variations in the force of the muscle contractions, weariness, ambient conditions, humidity, electrode movement, as well as other within-/between-day changes can all cause variations in EMG signals. Despite giving users functional dexterity, traditional myoelectric prosthesis control algorithms do not yet enable physiologically normal upper-limb movements.

Since the 1970s, upper-body part prosthetic scientists have been looking into the use of EMG signal-actuated machine learning methods to enable more flexible and intuitive myoelectric device control. Due in major part to developments in signal processing techniques, cognitively capable processors, and improved battery technology, these algorithms have shown promise in terms of enhancing prosthesis control accuracy and user friendliness. However, due to their alleged lack of resilience, devices that use algorithms for machine learning are frequently not deployed in clinical settings. To get over this restriction, prosthesis researchers are still looking into different EMG signal-driven device control mechanisms. The majority of available commercially upper-body part myoelectric devices employ an open-circuit control technique, in which the device is unable to be responsive to its surroundings. However, due to their alleged lack of resilience, devices that use machine learning programs are frequently not deployed in clinical settings. To get over this restriction, prosthesis researchers are still looking into different EMG signal-driven device control methods. The majority of upper-body part myoelectric devices that are commercially available employ open-loop control techniques, in which the device is not given feedback from its surroundings. Instead, after purposefully tightening a muscle in the remaining limb to begin and maintain control of the device's movement, a user is left to rely solely on optical feedback. Raw EMG signals are produced as a result of the electrical potentials created by this contraction. Typically, these signals are altered at a steady alteration between 200 and 1,000 Hz (depending on the category of EMG electrodes attached in the prosthetic socket). After that, the raw signals are analyzed, which involves cleaning and feature extraction. In order to make upper-limb prosthesis controls more "intelligent," or able to anticipate users' intended actions, machine learning is applied in the design of these controls. Researchers working on myoelectric prostheses presently employ a variety of machine learning techniques (based on concepts from statistics and computer science) to create more user-friendly device controls. Each technique calls for the creation of a systematic model that may be implemented to foretell the prescribed device movements using EMG signals recorded while wearing a prosthesis. As a result, a model acts as a tracing function that can convert Neuromuscular input signals from electrodes to instructions for a device's motor.

10 Conclusion

This chapter gives us an overview to the design and fabrication advances in prosthetics and orthotics. The different steps involved in fabricating a prosthetic such as creation of the three-dimensional model to the rapid prototyping operation of actually making the prosthetic part for the patient have been highlighted here. The different materials employed for making orthotics and prosthetics have also been discussed. Towards the end of the chapter, new age prosthetics such as neural prosthetics have been covered, and how they will affect patients with disability. Additionally, the application of machine learning in prosthetics has been touched briefly.

References

1. Wang Y, Tan Q, Pu F, Boone D, Zhang M (2020) A review of the application of additive manufacturing in prosthetic and orthotic clinics from a biomechanical perspective. Engineering 6(11):1258–1266. https://doi.org/10.1016/J.ENG.2020.07.019
2. Berke GM et al (2010) Comparison of satisfaction with current prosthetic care in veterans and servicemembers from Vietnam and OIF/OEF conflicts with major traumatic limb loss. J Rehabil Res Dev 47(4):361–372. https://doi.org/10.1682/JRRD.2009.12.0193
3. Totah D, Kovalenko I, Saez M, Barton K (2017) Manufacturing choices for ankle-foot orthoses: a multi-objective optimization. Procedia CIRP 65:145–150. https://doi.org/10.1016/J.PROCIR.2017.04.014
4. Yan Q et al (2018) A review of 3D printing technology for medical applications. Engineering 4(5):729–742. https://doi.org/10.1016/J.ENG.2018.07.021
5. Weller C, Kleer R, Piller FT (2015) Economic implications of 3D printing: market structure models in light of additive manufacturing revisited. Int J Prod Econ 164:43–56. https://doi.org/10.1016/J.IJPE.2015.02.020
6. Brier B, Vinh P, Schuster M, Mayforth H, Johnson Chapin E (2015) A radiologic study of an ancient egyptian mummy with a prosthetic toe. Anat Rec 298(6):1047–1058. https://doi.org/10.1002/AR.23135
7. Bisaccia M, Ao LM, Pio S, Rinonapoli G, Colleluori G (2016) The history of external fixation, a revolution idea for the treatment of limb's traumatized and deformities: from hippocrates to today supracondylar fractures of the humerus, Gartland 3: pediatric urgency. Case report and leterature review. View project c
8. Shahar FS et al (2019) A review on the orthotics and prosthetics and the potential of kenaf composites as alternative materials for ankle-foot orthosis. J Mech Behav Biomed Mater 99:169–185. https://doi.org/10.1016/J.JMBBM.2019.07.020
9. Prosthetics and Orthotics : Lower Limb and Spinal | Download file Author : Ron Seymour Pages : 540 pages Publisher : Lippincott Williams and Wilkins Language : English, p 13 (2002)
10. Au SK, Weber J, Herr H (2009) Powered ankle-foot prosthesis improves walking metabolic economy. IEEE Trans Robot 25(1):51–66. https://doi.org/10.1109/TRO.2008.2008747
11. Martinez-Villalpando EC, Mooney L, Elliott G, Herr H (2011) Antagonistic active knee prosthesis. A metabolic cost of walking comparison with a variable-damping prosthetic knee. Proc Annu Int Conf IEEE Eng Med Biol Soc EMBS, pp 8519–8522. https://doi.org/10.1109/IEMBS.2011.6092102
12. Chuang CH, Ko LW, Lin YP, Jung TP, Lin CT (2014) Independent component ensemble of EEG for brain-computer interface. IEEE Trans Neural Syst Rehabil Eng 22(2):230–238. https://doi.org/10.1109/TNSRE.2013.2293139

13. Tucker MR, Olivier J, Pagel A, Bleuler H (2015) Control strategies for active lower extremity prosthetics and orthotics: a review. J NeuroEngineering Rehabil 12:1. https://doi.org/10.1186/1743-0003-12-1
14. Tang L et al (2019) Functional gradient structural design of customized diabetic insoles. J Mech Behav Biomed Mater 94:279–287. https://doi.org/10.1016/J.JMBBM.2019.03.003
15. Diwakar M, Kumar M (2018) A review on CT image noise and its denoising. Biomed Signal Process Control 42:73–88. https://doi.org/10.1016/J.BSPC.2018.01.010
16. Haleem A, Javaid M (2019) 3D scanning applications in medical field: a literature-based review. Clin Epidemiol Glob Heal 7(2):199–210. https://doi.org/10.1016/J.CEGH.2018.05.006
17. Chua C, Meng CS, Ching LS, Teik LS, Aung SC (2000) Facial prosthetic model fabrication using rapid prototyping tools. Integr Manuf Syst 11(1):42–53. https://doi.org/10.1108/09576060010303668/FULL/PDF
18. Mavroidis C et al (2011) Patient specific ankle-foot orthoses using rapid prototyping. J Neuroeng Rehabil 8(1):1–11. https://doi.org/10.1186/1743-0003-8-1/FIGURES/10
19. Ciobanu O, Ciobanu G, Rotariu M (2013) Photogrammetric scanning technique and rapid prototyping used for prostheses and Ortheses Fabrication. Appl Mech Mater 371:230–234. https://doi.org/10.4028/WWW.SCIENTIFIC.NET/AMM.371.230
20. Lantada AD, Morgado PL (2012) Rapid prototyping for biomedical engineering: current capabilities and challenges 14:73–96. https://doi.org/10.1146/annurev-bioeng-071811-150112
21. Thompson MK et al (2016) Design for additive manufacturing: trends, opportunities, considerations, and constraints. CIRP Ann 65(2):737–760. https://doi.org/10.1016/J.CIRP.2016.05.004
22. Kruth JP (1991) Material incress manufacturing by rapid prototyping techniques. CIRP Ann 40(2):603–614. https://doi.org/10.1016/S0007-8506(07)61136-6
23. Paterson AM, Bibb R, Campbell RI, Bingham G (2015) Comparing additive manufacturing technologies for customized wrist splints. Rapid Prototyp. J. 21(3):230–243. https://doi.org/10.1108/RPJ-10-2013-0099/FULL/PDF
24. Herbert N, Simpson D, Spence WD, Ion W, A preliminary investigation into the development of 3-D printing of prosthetic sockets 42(2):141–146. https://doi.org/10.1682/JRRD.2004.08.0134
25. Schrank ES, Stanhope SJ (2011) Dimensional accuracy of ankle-foot orthoses constructed by rapid customization and manufacturing framework. J Rehabil Res Dev 48(1):31–42. https://doi.org/10.1682/JRRD.2009.12.0195
26. Deckers JP, Vermandel M, Geldhof J, Vasiliauskaite E, Forward M, Plasschaert F (2017) Development and clinical evaluation of laser-sintered ankle foot orthoses 47(1):42–46. https://doi.org/10.1080/14658011.2017.1413760
27. The 16th National Day on Biomedical Engineering
28. EBSCOhost | 17191301 | A preliminary investigation into the development of 3-D printing of prosthetic sockets
29. Saijo H et al. (2009) Maxillofacial reconstruction using custom-made artificial bones fabricated by inkjet printing technology. J Artif Organs 123, vol 12(3):200–205. https://doi.org/10.1007/S10047-009-0462-7
30. Lee Ventola C (2014) Medical applications for 3D printing: current and projected uses. Pharm Ther 39(10):704
31. Barrios-Muriel J, Romero-Sánchez F, Alonso-Sánchez FJ, Salgado DR (2020) Advances in orthotic and prosthetic manufacturing: a technology review. Materials (Basel) 13(2):295. https://doi.org/10.3390/MA13020295
32. Ali MH, Smagulov Z, Otepbergenov T (2021) Finite element analysis of the CFRP-based 3D printed ankle-foot orthosis. Procedia Comput Sci 179:55–62. https://doi.org/10.1016/J.PROCS.2020.12.008

33. Santhanam G, Ryu SI, Yu BM, Afshar A, Shenoy KV (2006) A high-performance brain–computer interface. Nat 4427099, 442(7099):195–198. https://doi.org/10.1038/nature04968
34. Wolpaw JR, McFarland DJ (2004) Control of a two-dimensional movement signal by a noninvasive brain-computer interface in humans. Proc Natl Acad Sci USA 101(51):17849–17854. https://doi.org/10.1073/PNAS.0403504101/SUPPL_FILE/03504MOVIE1.MOV
35. Ghafoor U, Kim S, Hong KS (2017) Selectivity and longevity of peripheral-nerve and machine interfaces: a review. Front Neurorobot 11:59. https://doi.org/10.3389/FNBOT.2017.00059/BIBTEX

Chapter 3
Research Progress of Self-Healing Elastomers Materials: Processing and Characterization

Viveksheel Rajput, Jasdeep Bhinder, and Gurpreet Singh

1 Introduction

Artificial or synthetically produced materials that have the innate capacity to spontaneously repair damage to themselves without the need for human involvement or external issue identification are known as self-healing materials [1, 2]. Due to its exceptional qualities, such as a high strength to weight ratio, fibre reinforcement composites are frequently used. The aerospace and automobile sectors may use these materials extensively. It has been observed that material failure ultimately results from fractures inside the material, which tend to change the material's mechanical, thermal, and electrical characteristics [3, 4]. Self-healing encompasses all kinds of materials, including metals, ceramics, and cementitious materials, even though polymers or elastomers are the most popular types of self-healing materials. The insertion of a repair agent that is housed in a tiny vessel or an inherent repair of the substance are two examples of different healing procedures. A substance must undergo the healing process unaided for it to be formally classified as autonomously self-healing. Nevertheless, self-healing polymers may start the healing process when exposed to an external stimulus (light, temperature change, etc.).

A self-healing mechanism can be obtained by embedding different healing agent carriers inside the composite such as hollow glass tubes [5], micro-encapsulation [6], and vascular network [7, 8]. For use in concrete, Dry [9–11] successfully created a self-healing structure based on the biological self-healing phenomenon known

V. Rajput (✉)
Sophisticated Analytical Instrumentation Facility (SAIF), Panjab University, Chandigarh, India
e-mail: sheelrajput03@gmail.com

J. Bhinder
École de Technologie Supérieure ÉTS, Montreal, Canada

G. Singh
Centre for Biomedical Engineering, Indian Institute of Technology Delhi, New Delhi, India

A. Chanda et al. (eds.), *Materials for Biomedical Simulation*, Materials Horizons: From Nature to Nanomaterials, https://doi.org/10.1007/978-981-99-5064-5_3

as bleeding. They spread the restorative chemicals throughout the concrete specimen after storing them within the containers. Once the vessels had been harmed by crack propagation, the repair agents had healed the cracks satisfactorily. There had been a considerable renovation of the buildings. Kasner et al. [12] indicates that the self-repairing abilities of living organisms have inspired many researchers to create lightweight materials with integrated healing mechanisms. It improves the service life of developed composite materials [13–15]. Li et al. [16] also developed self-repair method based on hollow glass tubes and applied crack repair mechanism in cement composite. Pang and Bond et al. [17] used 60 μm OD borosilicate glass tube, 50% hollow glass fiber fraction using a custom fiber fabrication technique. These HGFs are then incorporated together with carbon fiber or glass fiber into a polymer-based composite and mixed with resin without preservatives or curing agents to introduce self-healing performance into the composites. These fibers it fractures under load application and recovers its properties as it exits the damaged hollow glass tube into the damaged area, further preventing the damage from spreading. A significant increase in property restoration has been observed. Pang et al. [4] added a UV fluorescent dye with a healing agent to improve the visual effect of the repair mechanism.

Electrochemical Discharge machining (ECDM) process is utilized for machining the non-conductive materials such as glass, quartz etc. It applies the principle of thermal melting and chemical dissolution to remove or drill the material [18, 19]. As illustrated in Fig. 1, two electrodes are typically used in an ECDM process; one is the tool electrode and the other is the auxiliary electrode. Both electrodes are submerged in an electrolyte, which is typically an alkali substance such aqueous NaOH or KOH. To start electrolysis, a pulsed or continuous direct current supply is used to create a potential difference across the two electrodes (or to ignite electrochemical-reactions). Tiny gas bubbles of hydrogen and oxygen begin to develop at the cathode and anode respectively, as electrolysis begins to occur between the two electrodes. The ohmic heating and electrochemical reactions of the electrolyte cause the creation of these small bubbles. These little bubbles begin to combine to create a larger bubble. These small bubbles forms an insulating layer (or hydrogen film) of the gas bubbles close to the tool electrode as their nucleation density rises [20, 21]. This gas film, often referred to as tool blanketing, temporarily stops the passage of current while producing a strong electric field all around it. The electric breakdown of the film at a critical voltage (Vc), this accelerates the electrical spark generation across the tool and electrolyte. Rajput et al. [22], mentioned that electrolyte concentration has a major impact on electrolyte viscosity because its value rises with the increase in level of its concentration, producing smoother machined surfaces by boosting the electrolyte's chemical activity.

Kurafuji originally introduced ECDM in 1968 when micro-drilling holes in glass [23]. They claimed that specified drilling depths were the maximum for glass machining. A theoretical model for estimating the critical values of voltage and current necessary for spark production in ECDM was presented by Basak and Ghosh

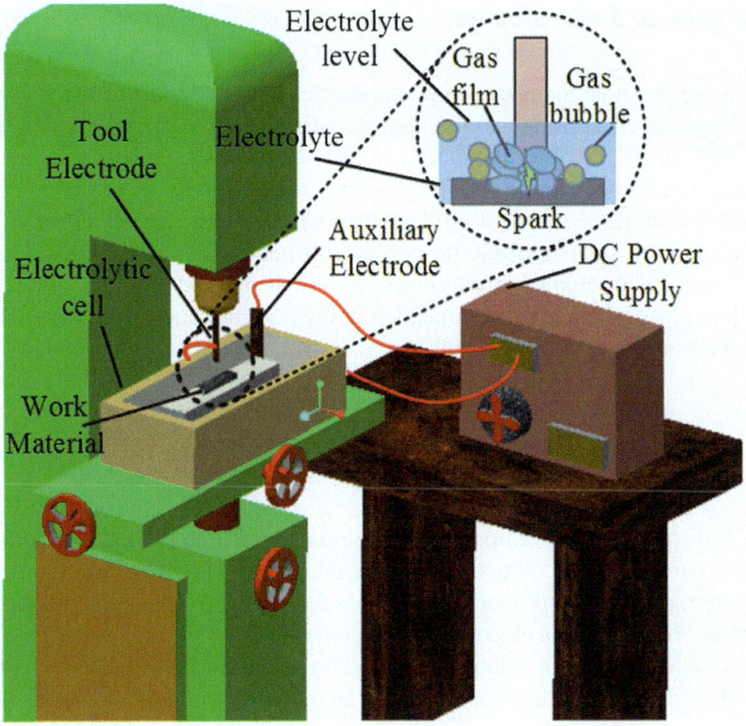

Fig. 1 Principle schematic diagram of ECDM process [19]

[24, 25]. Glass work materials were successfully machined by ECDM using efficient and accurate machining characteristics. Using ECDM during milling operations, Zheng et al. [26] created three-dimensional microstructures on glass material. Malik et al. [27] demonstrated that when non-conductive E-glass fibre composite was machined with NaOH electrolyte, the MRR of the workpiece increased in proportion to the applied voltage. Rajput et al. [28] mentioned that combining electrolyte with surfactant and raising its concentration can improve the chemical activity of the electrolytes. They found that the electrical conductivity of the electrolyte increased and as a result, better geometrical features are obtained. Various authors have assessed the impact of electrolyte concentration on the pace at which glass material was removed during the ECDM operation of micro-drilling [28–30]. They concluded that since there are more OH radicals present for etching action; electrolyte concentration considerably impacts the rate of material removal. Several authors have demonstrated and offered their explanations for how input process factors affect discrete response variables in ECDM process, and how multi-response optimization can be achieved to enhance its performance [31–34].

1.1 Problem Formulation

Literature reveals that many experimental studies have been performed to investigate the self-healing characteristics of the elastomers as well as the machining potential of the ECDM. So far, the machining of the self-healing elastomer using ECDM process has never been reported. The present study analyses the capability of ECDM process to machine the self-healing elastomer composites with ease and effectiveness. The healing performance of the developed self-healing elastomer is also assessed. Electrochemical discharge machining is one such method that can drill a hole in the hollow glass tubes that are embedded in the polymer composite that can further enable the pouring of healing agent.

2 Experimental Setup and Method

Figure 2 illustrates the development and integration of the developed adaptive based ECDM process into the vertical milling machine (VMC). It is based on the feedback system principle and uses adaptive tool feed control to carry out the necessary machining operations. An electrolytic cell constructed of polycarbonate material has a fixture for holding work materials. The performance and self-healing capability of the developed elastomer are analyzed using Impact strength. Subsequently, the developed self-healing elastomer is drilled using ECDM process to fabricate micro-hole in the HGTs embedded in the composites.

2.1 Adaptive Tool Feed

Adaptive tool feed based ECDM is utilised to obstruct the contact of the tool with the work material and to retain an efficient working space between them, in contrast to constant velocity tool feed and gravity tool feed and methods. Maintaining a

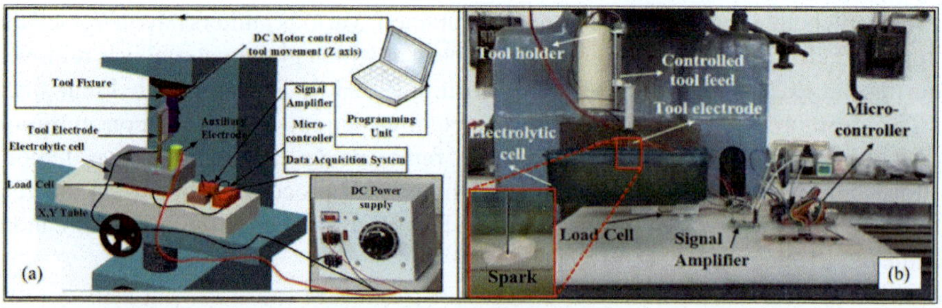

Fig. 2 **a** Cad model of the adaptive ECDM setup **b** Developed ECDM setup [19]

minimal separation without tool contact is referred to as an effective machining gap. It continually makes it possible for an efficient gas film to build beneath the tool electrode. Because there is constant tool contact with the work surface during a gravity tool feed, a gas film does not form under the tool electrode. The material below the tool is left unmachined or takes on a humped shape as a result of it only removing material from the sides of the tool [35]. Constant velocity tool feed has the drawback that the tool will touch the material surface if the MRR is lower than the feed rate of the tool. A significant machining gap is created under the tool if the MRR is greater than the feed rate of the tool. The heat energy from the sparks is lost into the electrolyte as a result [35, 36].

In this investigation, an adaptive tool feed control system that operates under real machining conditions is devised and constructed to address these issues. It adjusts the tool feed in accordance with the pace at which material is removed. It makes use of a sensitive load cell to track the moment when tool touches the material surface. The cell then generates the signal in the form of a potential difference as soon as the contact is detected, and this signal is then transmitted to the controller in an amplified way via an analogue to digital converter. The controller is configured to retract the tool upward while changing the stepper motor's orientation. The process capability or machinability of the developed setup is presented through drilled work materials as shown Fig. 3.

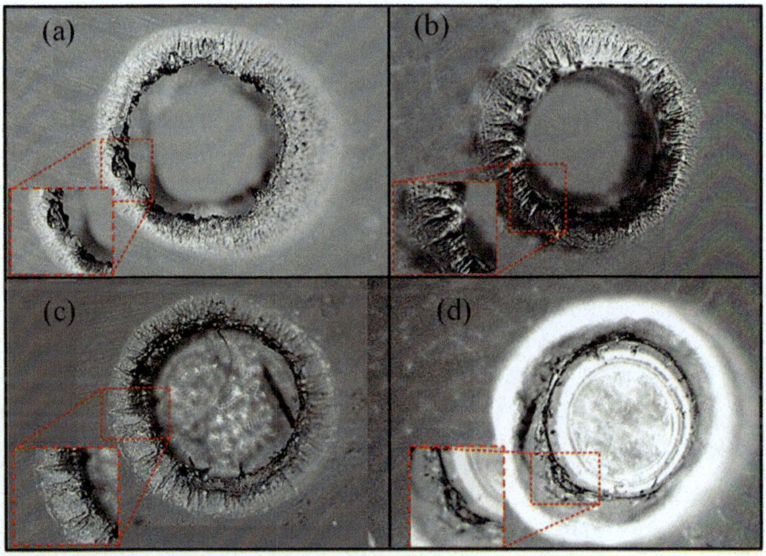

Fig. 3 Microscopy images of the work materials drilled using developed adaptive ECDM process [35]

3 Enhancing Visibility

A dye is included into the healing agent contained inside the HGTs and implanted throughout the composite to improve the visibility of the injury. The HGTs are filled by spring-assisted capillary action. Via bleeding under crack, the dye improves the visual inspection of the fracture and points in the direction of the healing agent's flow. Figure 4a shows the damage detection and healing agent release while Fig. 4b shows the fabricated self-healing elastomer having dye mixed healing agent within the HGTs. It not only reveals the composite's damaged region, but also the flow of dye-mixed healing agent into that area. After the initial damage causes a crack to propagate, it is found that the HGTs release a healing agent, supporting the self-healing strategy depicted in present study.

4 Results and Discussions

4.1 Self-Healing Assessment

An experimental investigation of the impact strength is performed to assess the self-healing capability of the developed elastomer. Figure 5 shows the damaged self-healing elastomer after the impact test. Figure 6 presents the comparison of the Impact strength of the elastomer with and without healing after the damage. An initial damage is given to the elastomer to initiate the crack propagation and flow of the healing agent. An increase of 26.43% in impact strength is observed when the elastomer is healed for 1 week. When elastomer is not given a healing period, it displays no restoration of its mechanical characteristics.

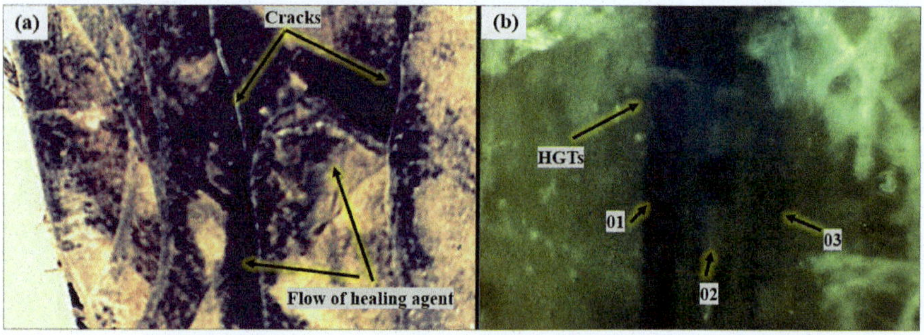

Fig. 4 **a** Flow of the healing agent after crack initiation **b** Dye mixed healing agent filled in HGTs embedded inside the developed self-healing elastomer

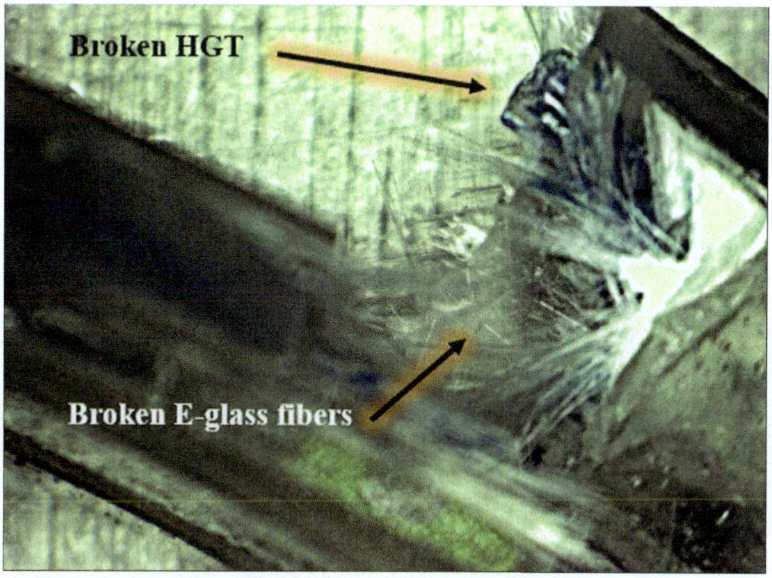

Fig. 5 Damaged Self-Healing elastomer

Fig. 6 Impact strength comparison of healed and not healed elastomer

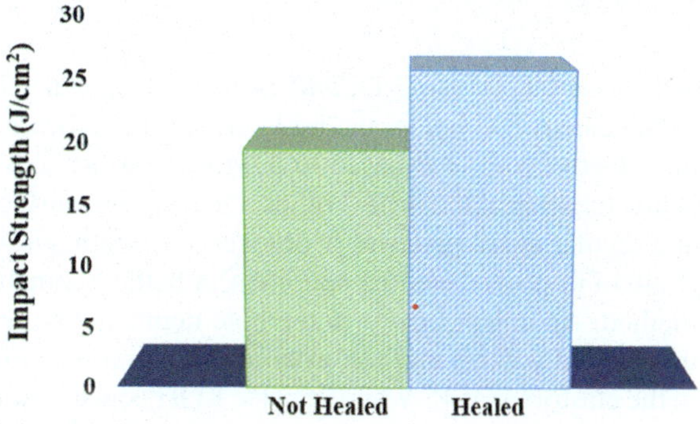

4.2 Electrochemical Discharge Based Drilling

The micro-holes are drilled on the self-healing material under varied machining circumstances. As a tool, a stainless-steel cylindrically shaped with a diameter of 1000 mm is used. As an electrolyte, potassium hydroxide (KOH) is employed. Table 1 provides the machining specifications for drilling the hole in the self-healing elastomer.

The condition of the holes in the self-healing elastomer is analyzed from the side of the tool entry by using FESEM image as shown in Fig. 7. The hole drilled at 35 V and 20 wt% KOH concentration. Results indicate that the holes can be successfully

Table 1 Machining conditions

Variable	Values	Variable	Values
Voltage	35–45 V	Electrode feed rate	4 mm/min
Concentration of the electrolyte	15–25% wt/v	Electrolyte	KOH
Inter electrode gap (IEG)	35 mm	Anode material	Graphite
		Tool material	Stainless steel
Electrolyte level	1 mm (approx.)	Machining time	2 min

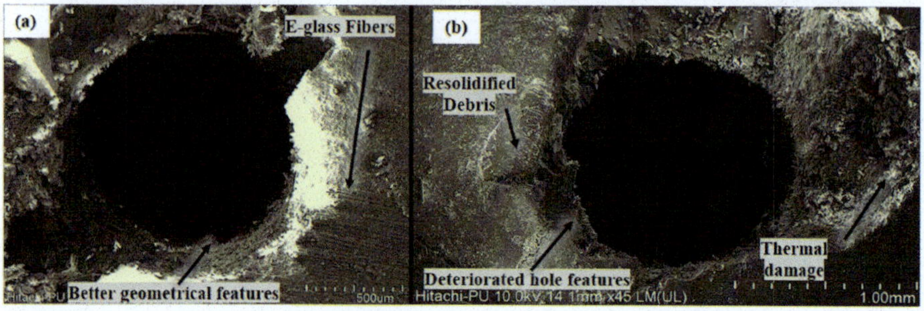

Fig. 7 FESEM images of drilled hole in HGTs in self-healing elastomers **a** 35 V and 20 wt% **b** 45 V and 25 wt%

drilled on HGTs using ECDM process is shown in Fig. 7a. Due to the presence of the electrolyte below the tool surface, the efficient machining gap is attained in the adaptive tool feed based ECDM that further aids in improving the geometrical characteristics of the drilled holes. On the work surface, it creates a uniform gas film and regular spark patterns. Moreover, a smooth surface is created at the edges as a result of improved etching activities. A better geometric profile is achieved with this method. An enhancement in the hole depth can be seen with the increment in both the applied voltage and electrolyte concentrations. Figure 7b shows the drilled hole in the elastomer at 45 V and 25 wt% KOH concentration. It is observed that the rapid formation of the gas film is achieved as a result of an improvement in the rate of bubble formation brought about by an increase in the applied voltage and electrolyte concentration. As a result, high spark frequencies are therefore obtained, increasing the thermal energy transmission. The sparking rate is accelerated by an enhancement in electrolyte conductivity with an increment in electrolyte concentration [11, 13]. Thus, a higher MRR is attained. However, a high number of thermal cracks and debris formation is observed at higher level of input variables.

5 Conclusions

The present study incorporates the experimental examination for the drilling of the self-healing elastomer composites using ECDM process. The self-healing capability of the developed elastomer is also analyzed using Impact test. The fabricated elastomer is provided with initial damage to produce the crack that further releases the healing agent. In the elastomer composite, a dye is added to improve both the visibility of the damage and the agent flow. The following is a list of the main conclusions reached in this study:

- ECDM process successfully produces the holes on the HGTs embedded inside the self-healing elastomer.
- An effective material removal rate is achieved using adaptive tool feed based ECDM during the hole fabrication in elastomer.
- Any increment in the electrolyte concentration and applied voltage leads to the increase in hole depth and material removal. However, higher thermal cracks are also observed due to high thermal energy.
- In self-healing elastomers, the use of hollow glass fibres (HGFs) to transport a healing agent has shown to be an effective technique.
- Due to crack filling, healed material show improved property restoration and increased self-healing. After the material is healed for 1 week, its flexural strength increases by 26.43%.
- Using a dye-mixed healing agent can improve the visibility of the damaged areas and the release of the healing agent.

6 Scope of Future Work

The future work can be carried out by the filling of healing agent at any time inside the HGTs once holes are drilled using ECDM process. Interestingly, it also compensates for the earlier reported 'fabrication error' of the composite material like partial filling of the tubes, crack development etc. Also, to provide a longer useful life for the self-healing elastomers, the amount of healing agent within the hollow glass fibres can also be optimised. To further improve the efficiency of ECDM machining, it is important to thoroughly investigate the effects of various electrochemical characteristics of the electrolytes (mixed electrolytes, surfactant mixed electrolytes, and abrasive mixed electrolytes) on the bubble amalgamation and bubble departure phenomena.

Acknowledgements The authors acknowledge the support and assistance given by the SAIF/CIL, Panjab University Chandigarh, India.

References

1. Kessler MR (2012) Polymer matrix composites: a perspective for a special issue of polymer reviews. Polym Rev 52(3):229–233
2. Bleay SM, Loader CB, Hawyes VJ, Humberstone L, Curtis PT (2001) A smart repair system for polymer matrix composites. Compos A Appl Sci Manuf 32(12):1767–1776
3. Brown EN, Kessler MR, Sottos NR, White SR (2003) In situ poly (urea-formaldehyde) microencapsulation of dicyclopentadiene. J Microencapsul 20(6):719–730
4. Pang JW, Bond IP (2005) A hollow fibre reinforced polymer composite encompassing self-healing and enhanced damage visibility. Compos Sci Technol 65(11–12):1791–1799
5. Trask RS, Bond IP (2006) Biomimetic self-healing of advanced composite structures using hollow glass fibres. Smart Mater Struct 15(3):704
6. Xue C, Li W, Li J, Tam VW, Ye G (2019) A review study on encapsulation-based self-healing for cementitious materials. Struct Concr 20(1):198–212
7. Hansen CJ, Wu W, Toohey KS, Sottos NR, White SR, Lewis JA (2009) Self-healing materials with interpenetrating microvascular networks. Adv Mater 21(41):4143–4147
8. Shields Y, De Belie N, Jefferson A, Van Tittelboom K (2021) A review of vascular networks for self-healing applications. Smart Mater Struct 30(6):063001
9. Dry C (1994) Matrix cracking repair and filling using active and passive modes for smart timed release of chemicals from fibers into cement matrices. Smart Mater Struct 3(2):118
10. Dry C, McMillan W (1996) Three-part methyl methacrylate adhesive system as an internal delivery system for smart responsive concrete. Smart Mater Struct 5(3):297
11. Dry C (1990) Alteration of matrix permeability, pore and crack structure by the time release of internal chemicals. In: Proceedings of the ACS/NIST Conference on Advances in Cementitious Material, American Ceramic Society, co-sponsored by National Institute of Standards and Technology, Gaithersburg, Maryland, vol 768
12. Kassner ME, Nemat-Nasser S, Suo Z, Bao G, Barbour JC, Brinson LC, Van Swol F (2005) New directions in mechanics. Mech Mater 37(2–3):231–259
13. Li VC, Lim YM, Chan YW (1998) Feasibility study of a passive smart self-healing cementitious composite. Compos B Eng 29(6):819–827
14. Rajput V, Goud M, Samir S (2017) A review of developments in the self-healing approaches of composite materials. Int J Mech Eng Technol 8:1699–1709
15. Nagori I, Goud M, Rajput V (2019) Self-healing composite materials: a review on preceding and perspective research. Int J Tec Innov Mod Eng Sci 5(5):687–697
16. Rong H, Wang M, Zhang Y, Lu X (2023) A high strength, high toughness and transparent room-temperature self-healing elastomer based on the synergy effect of quadruple dynamic bonding structure. React Funct Polym 185:105531
17. Pang JWC, Bond IP (2004) Self-Repair and enhanced damage visibility in a hollow fibre reinforced plastic. In: Proceedings 11th European Confernce on Composite Materials, Rhodes, May–June 2004
18. Fascio V, Wuthrich R, Viquerat D, Langen H (1999) 3D microstructuring of glass using electrochemical discharge machining (ECDM). In: MHS'99. Proceedings of 1999 International Symposium on Micromechatronics and Human Science (Cat. No. 99TH8478), IEEE, pp 179–183
19. Rajput V, Pundir SS, Goud M, Suri NM (2021) Multi-response optimization of ECDM parameters for silica (quartz) using grey relational analysis. SILICON 13:1619–1640. https://doi.org/10.1007/s12633-020-00538-7
20. (2023) Enhancement of Electrochemical Discharge Machining (ECDM) Characteristics with Tool Electrode Rotation. In: Advances in Modelling and Optimization of Manufacturing and Industrial Systems: Select Proceedings of CIMS 2021, Springer Nature Singapore, Singapore, pp 135–148
21. Bhargav KVJ, Balaji PS, Sahu RK, Katiyar JK (2023) Exemplary approach using tool rotation-assisted μ-ECDM for CFRP composites machining. Mater Manuf Processes 38(3):271–283

22. Rajput V, Goud M, Suri NM (2021) Electrochemical discharge machining: gas film electrochemical aspects, stability parameters, and research work. J Electrochem Soc 168(1):013503. https://doi.org/10.1149/1945-7111/abd516
23. Kurafuji H, Suda K (1965) Electrical discharge drilling of glass. CIRP Ann 16:415–419
24. Basak I, Ghosh A (1997) Mechanism of material removal in electrochemical discharge machining: a theoretical model and experimental verification. J Mater Process Technol 71(3):350–359
25. Basak I, Ghosh A (1996) Mechanism of spark generation during electrochemical discharge machining: a theoretical model and experimental verification. J Mater Process Technol 62(1–3):46–53
26. Zheng ZP, Cheng WH, Huang FY, Yan BH (2007) 3D microstructuring of Pyrex glass using the electrochemical discharge machining process. J Micromech Microeng 17(5):960
27. Malik A, Manna A (2016) An experimental investigation on developed WECSM during micro slicing of e-glass fibre epoxy composite. Int J Adv Manuf Technol 85:2097–2106
28. Rajput V, Goud M, Suri NM (2020) Performance analysis of ECDM process using surfactant mixed electrolyte. In: Manufacturing engineering: select proceedings of CPIE 2019. Springer, Singapore, pp 285–300. https://doi.org/10.1007/978-981-15-4619-8_22
29. Chikane P, Yadav P (2023) the role of electrolyte in ECDM process: a review. In: Li X et al (eds) Emerging trends in mechanical and industrial engineering: select proceedings of ICETMIE 2022. Springer, Singapore, pp 759–769
30. Rajput V, Goud M, Suri NM (2021) Review on recent advances, research trends, and gas film in electrochemical discharge-based micromachining. J Micro Nanomanuf. https://doi.org/10.1115/1.4049418
31. Prajapati M, Lalwani D (2023) Study on evolution of hydrogen gas bubble on the performance of the μ-ECDM process. In: Vasudevan H et al (eds) Proceedings of International Conference on Intelligent Manufacturing and Automation: ICIMA 2022. Springer Nature Singapore, Singapore, pp 281–292
32. Rajput V, Goud M, Suri NM (2021) Three-dimensional finite element modeling and response surface based multi-response optimization during silica drilling with closed-loop ECDM. SILICON. https://doi.org/10.1007/978-981-15-4619-8_22
33. Rajput V, Goud M, Suri NM, Singh S, Chaudhary GR (2022) Surface characteristics study of Ti–6Al–4V alloy for biomedical applications using finite element modeling. J Mater Res 37(16):2710–2721. https://doi.org/10.1557/s43578-022-00677-0
34. Kumar M, Vaishya RO, Suri NM (2022) Simulation and experimental research of material removal rate in micro-electrochemical discharge machining process. Adv Mater Process Technol. https://doi.org/10.1080/2374068X.2022.2127936
35. AbouZiki JD, Wüthrich R (2012) Tool wear and tool thermal expansion during micromachining by spark assisted chemical engraving. Int J Adv Manuf Technol 61:481–486
36. Rajput V, Goud M, Suri NM (2022) Performance analysis of closed-loop electrochemical discharge machining (CLECDM) during micro-drilling and response surface methodology based multi-response parametric optimisation. Adv Mater Process Technol 8(2):1352–1382. https://doi.org/10.1080/2374068X.2020.1860494
37. Rajput V, Goud M, Suri NM (2022) 3D finite element modeling and multi-objective optimization for controlling the electrochemical discharge drilling parameters using the tool feed monitoring system. Proc Inst Mech Eng C J Mech Eng Sci 236(14):7829–7852. https://doi.org/10.1177/09544062221079170

Chapter 4
Biomechanical Modelling of Hierarchical Metamaterials for Skin Grafting

Vivek Gupta and Arnab Chanda

1 Introduction

The skin is mainly divided into three layers. The upper layer is called the epidermal layer, middle layer called dermal layer and bottom layer called hypodermal layer [1]. Skin performs several functions including preventing loss of moisture, reducing harmful effects of UV radiation and others [2]. There are several reasons which affect the skin properties such as electric stove heating, fires, accidents, cylinder explosions and many others [3]. The most common technique, which is used for burn injuries were full-thickness skin grafting (FTSG) and split-thickness skin grafting (STSG) [4–6].

Severe burns can be a devastating injury, often requiring extensive medical treatment to promote healing and prevent infection [3]. One crucial aspect of burn treatment is the use of skin grafts to cover damaged areas of skin [7]. However, due to the shortage of healthy skin, skin graft growth is a essential component of the healing process for large burn areas [8]. Skin grafting is the method of transplanting healthy skin from one region of the body to another to cover a wound or injury [2]. Full-thickness skin grafting (FTSG) and split-thickness skin grafting (STSG) are the two most common methods for skin grafting (STSG) [4–6]. In FTSG, the complete epidermis and dermis layer is removed from the abdomen. This technique is used for covering large burn areas and other wounds that require a thick, durable skin graft [9].

V. Gupta · A. Chanda (✉)
Centre for Biomedical Engineering, Indian Institute of Technology (IIT) Delhi, New Delhi, India
e-mail: arnab.chanda@cbme.iitd.ac.in

A. Chanda
Department of Biomedical Engineering, All India Institute of Medical Sciences (AIIMS) Delhi, New Delhi, India

A. Chanda et al. (eds.), *Materials for Biomedical Simulation*, Materials Horizons: From Nature to Nanomaterials, https://doi.org/10.1007/978-981-99-5064-5_4

STSG, on the other hand, includes removing only the epidermis and a part of the dermis from a healthy area of skin [9, 10]. This technique is often used for large burns due to its quicker healing time of 4–6 weeks [11, 12]. Additionally, STSG is frequently utilized for its ability to expand the skin graft by creating numerous parallel lines of small cuts on the intact removed skin using an appropriate skin grafting technique [2]. The skin grafting technique helps increase the capacity of STSG expansions, and the expansion is stated as a meshing ratio (MR) [13, 14]. Although several companies claim that STSGs can achieve skin expansions or meshing ratios of up to nine, clinical trials have shown that the highest expansion possible for STSG is just three [8, 15]. This emphasizes the importance of careful planning and management in skin grafting procedures, as well as the need for ongoing research to improve the effectiveness of burn treatments [16, 17].

Additive manufacturing (AM) techniques, mostly known as 3D printing, are being utilized to create skin graft patterns that can be tested using tensile testing machine. Gupta et al. [18] developed skin graft phantoms using AM techniques to determine the meshing ratios for low slit lengths and spacings. This allowed them to test and optimize skin graft designs before performing actual surgeries on patients. Javaid et al. [19] employed AM to design and create organ and scaffold components for tissue and organ printing. They found that 3D printing with scanned data could generate complex interior structures and could also be utilized to produce bone tissues for treating bone problems. Makode et al. [20] used two-part polymeric material, silicone, to fabricate skin simulant molds with matrix and collagen fiber oriented from 0° to 90°. By studying the variation in stress for several oriented skin simulants, they identified the best orientations for skin grafts. Baranski et al. [21] used a micropatterned polydimethylsiloxane (PDMS) template to align endothelial cells (ECs) within a collagen gel using AM. This allowed them to create complex vascular structures, which can be used to create clinically relevant heterogeneous tissue constructions. Vyas et al. [22] co-printed several cell-laden bio-inks to construct vasculature. This technique allows for the creation of complex and clinically relevant tissue structures.

Recently in 2022, Singh et al. [23] fabricated RT-shaped auxetic skin graft phantoms with varying angles of 0° to 135° using AM. In their study, they calculated stress, expansion, meshing ratio, and strain to identify the optimal design parameters for skin grafts. They concluded that skin grafts with a low RT angle would show maximum expansion. Overall, AM techniques are becoming an important tool in the design and testing of skin graft models. They allow for the creation of complex and customized structures that can be optimized for maximum effectiveness. By using these techniques, researchers can advance the field of skin grafting and improve outcomes for patients.

Wide-meshed skin grafts are often used to cover large areas of burn wounds and other skin injuries [24]. However, these grafts are usually weaker than normal skin due to the gaps in the mesh structure [25]. The mechanical strength of the graft depends on several factors, such as the type of mesh used and the technique used to attach the mesh to the wound bed [26]. To improve the mechanical strength of wide meshed skin grafts, various techniques have been developed [24, 27]. One of the approaches is to use an artificial dermis or dermal substitute to provide additional support for the mesh [28]. This can be achieved by placing a synthetic or biological material between the graft and the wound bed to promote tissue regeneration and healing [28]. Another approach is to use a modified mesh with a tighter structure or a different material composition to improve its strength. This modified mesh can be made of materials such as silicone or polyurethane, which are more durable and less prone to tearing [29]. In addition to these approaches, growth factors or other wound healing agents can be topically applied to the wound to improve tissue regeneration and wound healing. This can contribute to the overall strength of the graft [30].

There are two types of materials based on their Poisson's ratio: those with a positive Poisson's ratio and those with a negative Poisson's ratio [31]. The latter are referred to as auxetic materials, and they exhibit distinct characteristics from natural materials [32, 33]. Researchers have recently explored the use of auxetic patterns for designing skin grafts with better expansion potential [8, 23, 34]. For instance, Gupta et al. [34] investigated the effect of auxetic structures on the expansion potential of skin grafts and observed that these patterns had higher expansion than traditional ones. In a similar study, Gupta et al. [8] conducted a parametric analysis of various auxetic patterns with varying dimensional parameters and found that these structural changes could affect the axial change and expansion potential of the materials. Hierarchical auxetic structures with negative Poisson's ratio were found to exhibit higher strengths [35–37]. These structures are characterized by substructures with their geometry, allowing them to undergo multi-level deformation processes and exhibit unique mechanical behaviors. This approach mimics the natural environment to determine its role in strengthening materials and has been applied in various fields, including construction cranes and scaffolding [35–37].

The current study focuses on the development of hierarchical auxetic skin graft simulants, which can potentially improve burn surgery outcomes. The authors designed alternating slit (AS) and rotating rectangle (RR) shaped auxetic skin grafts using a design tool, and additive manufacturing was used to develop the molds. Silicone was then used to fabricate the hierarchical auxetic structures, which were evaluated for their expansion potential. The study estimated various parameters such as stress–strain, Poisson's ratio, meshing ratio, and void area. The modeling and testing methodologies used for skin graft simulants are discussed in Sect. 2, and the results are presented in Sect. 3, followed by the conclusions in Sect. 4. Overall, the study highlights the potential of hierarchical auxetic skin grafts as a promising approach for improving burn surgery outcomes.

2 Materials and Methods

2.1 Geometrical Modeling

CAD modelling was used to design the hierarchical patterns. The dimensions and schematic of all the designs was shown in Table 1 and Fig. 1 respectively. The dimensions of the designs were selected from the prior studies. Total four CAD models was designed using the 3D modelling software called SolidWorks (Dassault Systèmes, Vélizy-Villacoublay, France). For this study, the outer dimensions of the hierarchical patters were 50 mm × 50 mm × 2 mm. Figure 1 illustrates the first- and second-order AS and RR-shaped auxetic skin graft models. For the design of the first-order AS-shaped auxetic skin graft, the length of the alternating slits was retained the same as in earlier studies of AS-shaped and its derivatives. To investigate the influence of dimension parameters, the length of alternate slits in the first-order RR-shaped design was not equal. The number of slits was same in first-order AS and RR-shaped auxetic designs. The auxetic structures of the second-order hierarchy had more slits.

2.2 Fabrication of Skin Graft Simulants

The section outlines a method for producing a substance that efficiently mimics the epidermis and dermis at a depth of 2 mm (Fig. 2). The materials composition was produced applying a polymeric substance that has been utilized in comparable investigations on cadaveric skin and skin simulants. Several skin simulant pieces were produced and evaluated employing uniaxial stress at a 24 mm/min displacement rate, similar to previous studies [20, 38–41]. These compositions mechanical characteristics were compared to the mechanical properties of cadaver skin [42]. On the basis of the results of these experiments, an appropriate skin simulant composition was chosen. A polymeric material with a shore hardness of 5A was mixed with another polymeric material having a shore hardness of 30A in a weight ratio of 1:1. Pouring the compound into moulds with hierarchical auxetic patterns. The mixture was kept for 6 to 8 h to cure. The ultimate shore hardness of the skin graft simulants was 15A ± 2A, which was comparable to the qualities of cadaveric skin. In conclusion, the section presents a method for generating a polymeric composition

Table 1 Hierarchical skin graft model parameters, in millimetres

AS 1st order	H_1 or $L_1 = 17.92$	$H'_1 = 8.96$	$H''_1 = 4.48$	$T_{H1} = 11.2$
AS 2nd order	H_2 or $L_2 = 8.96$	$H'_2 = 4.48$	$H''_2 = 2.24$	$T_{H2} = 5.6$
RR 1st order	$X_1 = 8.96$	$L'_1 = 4.48$	$L''_1 = 4.48$	$T_{L1} = 11.2$
RR 2nd order	$X_2 = 4.48$	$L'_2 = 2.24$	$L''_2 = 2.24$	$T_{L2} = 5.6$

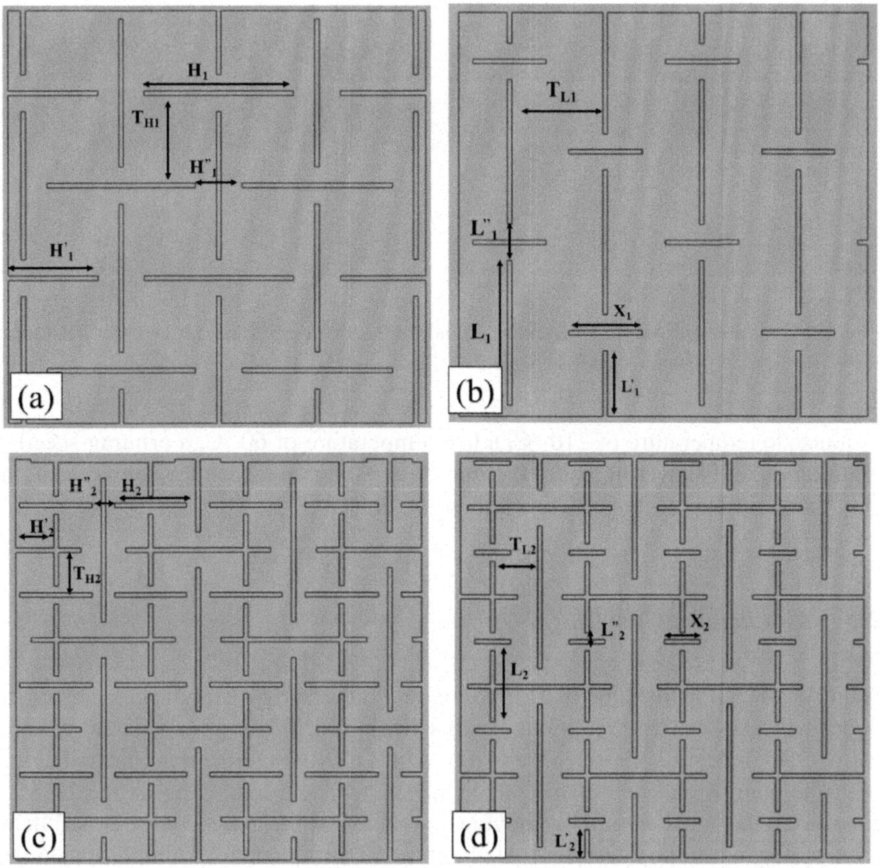

Fig. 1 Hierarchical auxetic skin graft designs **a** 1st order AS-shaped **b** 1st order RR-shaped **c** 2nd order AS-shaped **d** 2nd order RR-shaped

that replicates the the epidermis and 2 mm of dermis at medium depth. This composition was produced by repeated testing and comparison to cadaveric human skin, yielding a composition with a shore hardness comparable to that of cadaveric skin. After molding and curing the material, skin graft simulants were developed.

The process of developing 3D moulds for hierarchical skin grafts using additive manufacturing techniques. The dimensions of the moulds were consistent at 70 mm × 50 mm × 2 mm with an offset of 2 mm. The design and development of four skin graft moulds, each with a hierarchy of order up to two, was accomplished. Additive manufacturing process was used to process a 3D model, which is created using a computer-aided design (CAD) software, and then a 3D printer is used to print the model layer by layer [43–46]. In this case, the STL (STereoLithography) files of the moulds were converted into g-code and printed using a 3D printer that used Polylactic Acid (PLA) as a printing material. The printer was run with certain parameters, such

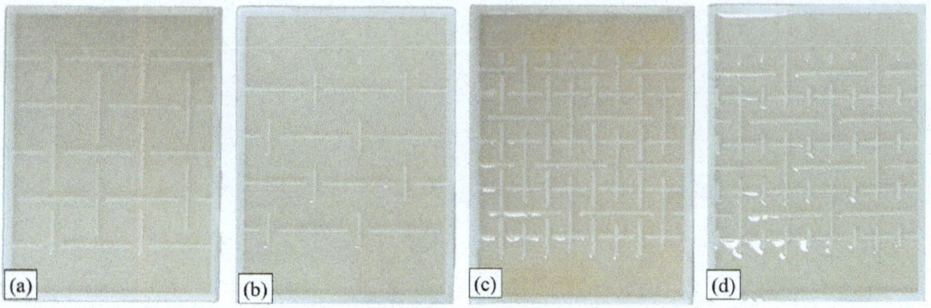

Fig. 2 Skin grafts with varying hierarchical moulds **a** 1st order AS-shaped **b** 1st order RR-shaped **c** 2nd order AS-shaped **d** 2nd order RR-shaped

as a nozzle temperature of 210 °C, a bed temperature of 60 °C, a printing speed of 45 mm/sec, and a layer height of 0.1 mm. When the moulds had been produced, the skin graft simulants were cast using these moulds.

2.3 *Mechanical Testing*

A universal testing machine (UTM) commonly utilized for material characterization under uniaxial loading conditions. In this work, four hierarchical auxetic patterns were tested under uniaxial tensile condition. In UTM, the sample was attached tightly with the clamps and one clamp was attached with the load cell. As the UTM is functional, the upper clamp was moving and sample stretched. All these experiments were performed at constant speed of 0.4 mm/sec. The schematic of sample attachment and load cell was shown in Fig. 3. The force–displacement data was converted into stress(σ)-strain (ε) data using the Eq. 1. The Poisson's ratio of all thc hicrarchical patterns was calculated using Eq. 2, under different stretching conditions. However, it was noted that the effective Poisson's ratio changes with the type of materials used. To determine the longitudinal strain, the maximum displacement that could be delivered in the direction of the uniaxial loading was divided by the hierarchical skin graft simulant initial length (L). Using Eq. 3, we were able to determine the lateral strain by dividing the maximal orthogonal displacement by the width of the graft model at the beginning of the calculation. The meshing ratio (MR) was measured during the uniaxial testing to determine the expansion of the skin graft simulants. The MR was defined as the expanded area of the skin graft to the unexpanded area of the skin graft, using Eq. 4. Void area is crucial parameters for cellular proliferation in developing cells. Void area. In this skin grafting work, the void area, which refers to an area that is empty or contains nothing, is extremely crucial. Using imaging methods, the empty area and maximum edge length were computed by determining the maximum value of X and Y. In summary, the methods used to perform uniaxial testing on hierarchical auxetic skin graft simulants using a UTM. The force–displacement data

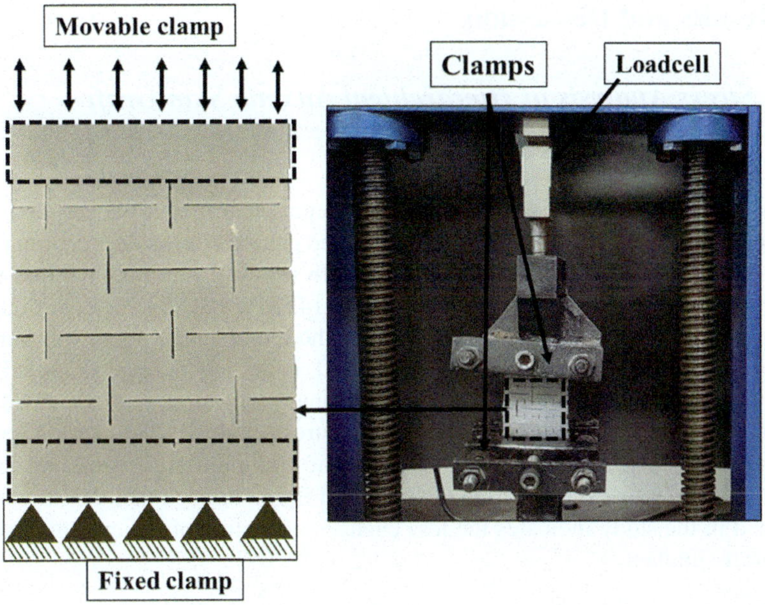

Fig. 3 Uniaxial testing configuration for skin graft simulators on the UTM

obtained from the UTM were converted to stress–strain. The effective Poisson's ratio, longitudinal strain, and lateral strain were estimated using Eqs. 1–3. The expansion of the skin graft simulants was measured using the meshing ratio, as defined by Eq. 4.

$$\text{Stress } (\sigma) = \frac{\text{Force (F)}}{\text{Cross}-\text{section area (A)}}, \tag{1}$$

$$\text{Poissons ratio } (\nu_{12}) = -\frac{d\varepsilon_2}{d\varepsilon_1}, \tag{2}$$

where $d\varepsilon_1, d\varepsilon_2$ are the longitudinal and lateral strain.

$$d\varepsilon_1 = \frac{dL_1}{L_1} \textit{ and } d\varepsilon_2 = \frac{dL_2}{L_2}, \tag{3}$$

$$\text{MR} = \frac{\text{Expanded Area}}{\text{Unexpanded Area}}. \tag{4}$$

3 Results and Discussion

3.1 Stress Analysis of Hierarchical Auxetic Skin Graft Simulants

Figure 4 illustrates the stress–strain values of hierarchical auxetic patterns at different stretching levels up to 300%. 1st order RR shaped patterns shows the maximum stress values and 2nd order AS shaped auxetic pattern shows the minimum stress values. The maximum and minimum stress value was 145 kPa and 85 kPa. It was observed that up to 20% strain, the stress values in all the hierarchical model was very low. Similar work conducted by the Gupta et al. [18], in their work, oval shape skin graft patterns was fabricated with varying dimensional parameters and calculated stress values were under the same range. From the stress analysis, 2nd order AS shaped hierarchical auxetic skin graft simulant shows the minimum stress value and could be best combination for skin grafting technique. It is emphasized that the lowest stress obtained in the study indicates the less chances of rupture of a hierarchical auxetic skin graft simulant.

3.2 Poisson's Ratio Analysis of Hierarchical Auxetic Skin Graft Simulants

The results of Poisson's ratio analysis for hierarchical auxetic skin graft simulants, as shown in Fig. 5. The Poisson's ratio values were calculated at 100%, 200%, and 300% strain. The negative Poisson's ratio values decreased for all skin graft models from 100 to 300% strain. At 100% strain, the 2nd order AS-shaped and 1st order RR-shaped auxetic skin graft simulant showed the highest and lowest negative Poisson's ratios (− 1.7 and − 1.5), respectively. Similarly, at 200% and 300% strains, the 2nd

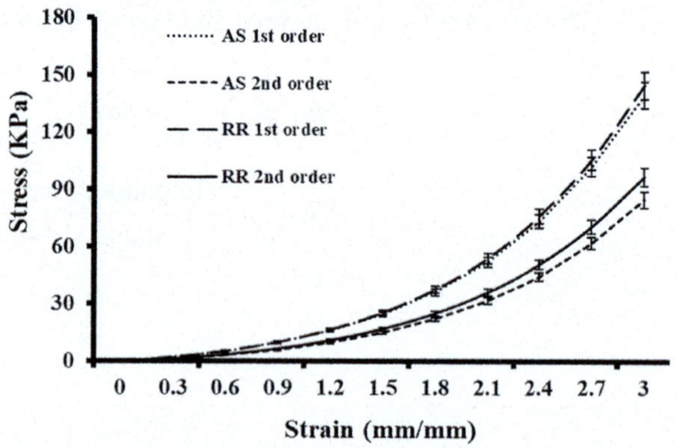

Fig. 4 Stress–strain plot of hierarchical auxetic skin graft simulants up to 300% strain

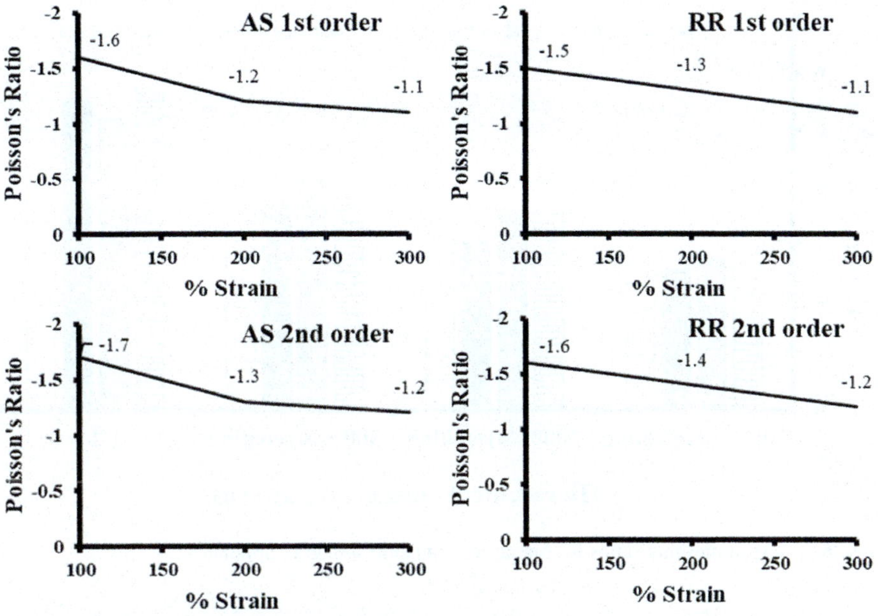

Fig. 5 Poisson's ratio of hierarchical auxetic skin graft simulants

order AS-shaped auxetic skin graft simulant and the 1st order RR-shaped auxetic skin graft simulant exhibited the highest and lowest negative Poisson's ratios, respectively. Overall, the hierarchical structures in this study showed negative Poisson's ratio values, and the 2nd order AS-shaped auxetic structure exhibited the highest Poisson's ratio. Based on these results, this structure may be the most suitable skin graft structure for covering large burn areas.

3.3 *Meshing Ratio Analysis of Hierarchical Auxetic Skin Graft Simulants*

Figure 6 illustrates the meshing ratio (MR) of the hierarchical auxetic skin graft simulants up to their ultimate tensile strength (UTS). The study conducted five tests on each of the skin graft models to ensure repeatability of the data. The results indicated that stretching caused an increase in the MR values of the hierarchical structures. At 100%, 200%, 300% strain, and UTS elongation, the 2nd order RR-shaped skin graft simulant exhibited the lowest MR values, whereas the 1st order AS-shaped patterns showed the highest MR values. The 1st order AS-shaped auxetic skin graft simulant had a maximum MR value of 5.8, while the 2nd order RR-shaped had the minimum MR value of 5 at UTS. The other two hierarchical skin graft simulants, namely the 2nd order AS-shaped and the 1st order RR-shaped, showed

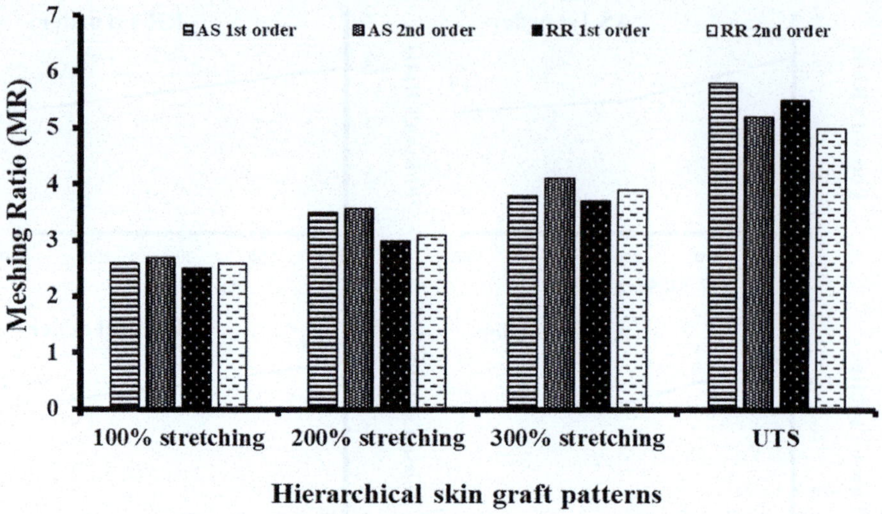

Fig. 6 Uniaxial meshing ratios for hierarchical auxetic skin graft simulants

MR values of 5.2 and 5.5, respectively. Therefore, the skin graft covering the largest area with cell growth and other clinical aspects will be the most appropriate for skin grafting applications and will cover extensive burn areas. The meshing ratio value can provide insights into the strength and structural integrity of the skin graft, which is crucial for ensuring its success in covering and protecting burn areas.

3.4 Void Area Analysis of Hierarchical Auxetic Skin Graft Simulants

The void area of hierarchical auxetic skin graft patters of each unit cell at different levels was shown in Table 2. The void area, which represents the space between the unit cells, is a crucial parameter affecting the skin ability to regenerate and heal. Higher void areas can impede cell proliferation, leading to reduced graft success rates and poor wound healing. However, biological investigations, such as studies with cadaveric and animal skins and healing agents, are needed to estimate cell proliferation and wound healing in these designs. Overall, the 1st order RR-shaped skin graft simulants showed a higher void area compared to the all other skin graft models. 2nd order AS-shaped skin graft simulants shows the minimum void area. Overall The results suggest that the 2nd order AS-shaped hierarchical auxetic skin graft designs coude be the best possible graft, which is used in skin grafting applications.

Table 2 Void area values of different hierarchical patterns

Design	Initial void area (mm^2)	Void area at 100% (mm^2)	Void area at 200% (mm^2)	Void area at 300% (mm^2)
AS 1st order	35.84	167	260	427
AS 2nd order	35.84	186	243	392
RR 1st order	35.84	183	226	463
RR 2nd order	35.84	181	221	459

4 Conclusions

This work details an experimental investigation into the biomechanics of hierarchical skin graft simulants. To develop skin graft simulants, AS- and RR-shaped auxetic patterns with distinct levels of hierarchy were developed. 3D printing techniques were utilized to manufacture moulds for the biofidelic material that replicates the mechanical properties of human skin. Uniaxial tensile experiments were conducted to evaluate the stress caused by hierarchical structures on skin graft simulants. Poisson's ratio, meshing ratio and void area was calculated to understand the maximum expansion and maximum void reign after stretching. At maximum tensile strength, the 2nd AS shaped auxetic skin graft simulant displayed the minimum indused stress, maximum Poisson's ratio and minimum void region, greatest meshing ratio (UTS). Overall, the 2nd order AS-shaped auxetic skin graft simulant was the most stable design for expansion without the risk of skin rupture. Also, this study provides insight into the biomechanics of hierarchical skin graft simulants and demonstrates the potential for expanding skin grafts using auxetic patterns. The findings suggest that such an approach may be a promising avenue for improving the outcomes of burn surgery.

References

1. Singh G, Chanda A (2021) Mechanical properties of whole-body soft human tissues: a review. Biomed Mater 16(6):062004. https://doi.org/10.1088/1748-605X/AC2B7A
2. Capek L, Flynn C, Molitor M, Chong S, Henys P (2018) Graft orientation influences meshing ratio. Burns 44(6):1439–1445. https://doi.org/10.1016/J.BURNS.2018.05.001
3. MacNeil S (2007) Progress and opportunities for tissue-engineered skin. Nature 445(7130):874–880. https://doi.org/10.1038/nature05664
4. Narayan N, Shivaiah R, Kumar KM (2021) A novel technique of collagen application over meshed split thickness graft for wound coverage. Int J Surg Med 7:54–61. https://doi.org/10.5455/ijsm.Collagen-Application-Meshed-Split-Thickness-Graft
5. Bogdanov SB, Gilevich IV, Melkonyan KI, Sotnichenko AS, Alekseenko SN, Porhanov VA (2021) Total full-thickness skin grafting for treating patients with extensive facial burn injury: a 10-year experience. Burns 47(6):1389–1398. https://doi.org/10.1016/j.burns.2020.12.003
6. Noureldin MA, Said TA, Makeen K, Kadry HM (2022) Comparative study between skin micrografting (Meek technique) and meshed skin grafts in paediatric burns. Burns 48(7):1632–1644. https://doi.org/10.1016/j.burns.2022.01.016

7. Hosseini M, Shafiee A (2021) Engineering bioactive scaffolds for skin regeneration. Small 17(41):2101384. https://doi.org/10.1002/smll.202101384
8. Gupta V, Chanda A (2022) Expansion potential of skin grafts with alternating slit based auxetic incisions. Forces Mech 7:100092. https://doi.org/10.1016/J.FINMEC.2022.100092
9. Claeys S (2017) Skin grafting. In: Griffon D, Hamaide A (eds) Complications in small animal surgery. StatPearls Publishing, Chichester, pp 561–568
10. Kagan RJ et al (2013) Surgical management of the burn wound and use of skin substitutes: an expert panel white paper. J Burn Care Res 34(2):e60–e79. https://doi.org/10.1097/BCR.0b013e31827039a6
11. Fan R, Hao R, McCarthy A, Xue J, Chen S (2023) Skin involved nanotechnology. In: Ning G (ed) Nanomedicine. Springer, Singapore, pp 719–753. https://doi.org/10.1007/978-981-16-8984-0_31
12. Zhou J, Scott C, Miab ZR, Lehmann C (2023) Current approaches for the treatment of ketamine-induced cystitis. Neurourol Urodyn 42(3):680–689. https://doi.org/10.1002/nau.25148
13. Huri D, Mankovits T (2018) Comparison of the material models in rubber finite element analysis. IOP Conf Ser Mater Sci Eng 393(1):012018. https://doi.org/10.1088/1757-899X/393/1/012018
14. Ozhathil DK, Tay MW, Wolf SE, Branski LK (2021) A narrative review of the history of skin grafting in burn care. Medicina 57(4):380. https://doi.org/10.3390/medicina57040380
15. Vandeput J, Nelissen M, Tanner JC, Boswick J (1995) A review of skin meshers. Burns 21(5):364–370. https://doi.org/10.1016/0305-4179(94)00008-5
16. Tang Y, Yin J (2017) Design of cut unit geometry in hierarchical kirigami-based auxetic metamaterials for high stretchability and compressibility. Extrem Mech Lett 12:77–85. https://doi.org/10.1016/J.EML.2016.07.005
17. Payne T, Mitchell S, Bibb R, Waters M (2014) Initial validation of a relaxed human soft tissue simulant for sports impact surrogates. Procedia Eng 72:533–538. https://doi.org/10.1016/j.proeng.2014.06.092
18. Gupta V, Chanda A (2022) Biomechanics of skin grafts: effect of pattern size, spacing and orientation. Eng. Res. Express 4(1):015006. https://doi.org/10.1088/2631-8695/AC48CB
19. Javaid M, Haleem A (2020) 3D printed tissue and organ using additive manufacturing: an overview. Clin Epidemiol Glob Heal 8(2):586–594. https://doi.org/10.1016/J.CEGH.2019.12.008
20. Makode S, Singh G, Chanda A (2021) Development of novel anisotropic skin simulants. Phys Scr 96(12):125019. https://doi.org/10.1088/1402-4896/AC2EFD
21. Baranski JD et al (2013) Geometric control of vascular networks to enhance engineered tissue integration and function. Proc Natl Acad Sci USA 110(19):7586–7591. https://doi.org/10.1073/PNAS.1217796110/SUPPL_FILE/PNAS.201217796SI.PDF
22. Vyas C, Pereira R, Huang B, Liu F, Wang W, Bartolo P (2017) Engineering the vasculature with additive manufacturing. Curr Opin Biomed Eng 2:1–13. https://doi.org/10.1016/j.cobme.2017.05.008
23. Singh G, Gupta V, Chanda A (2022) Mechanical characterization of rotating triangle shaped auxetic skin graft simulants. Facta Univ Ser Mech Eng. https://doi.org/10.22190/FUME220226038S
24. Shree A, Vagga AA (2022) Methodologies of autologous skin cell spray graft. Cureus. https://doi.org/10.7759/cureus.31353
25. Cigna E, Bolletta A, Giardino FR, Patanè L (2022) Grafts in plastic surgery. In: Maruccia M, Giudice G (eds) Textbook of plastic and reconstructive surgery. Springer, Cham, pp 61–75
26. Radharaman P, Kumar AKS, Kumar Sharma R (2019) The role of recruited minced skin grafting in improving the quality of healing at the donor site of split-thickness skin graft—A comparative study. Burns 45(4):923–928. https://doi.org/10.1016/j.burns.2018.11.018
27. Noor A et al (2022) Dressings for burn wound: a review. J Mater Sci 57(12):6536–6572. https://doi.org/10.1007/s10853-022-07056-4
28. Mandal BB, Kapoor S, Kundu SC (2009) Silk fibroin/polyacrylamide semi-interpenetrating network hydrogels for controlled drug release. Biomaterials 30(14):2826–2836. https://doi.org/10.1016/J.BIOMATERIALS.2009.01.040

29. Bell MA, Becker KP, Wood RJ (2022) Injection molding of soft robots. Adv Mater Technol 7(1):2100605. https://doi.org/10.1002/admt.202100605
30. Kaur G, Narayanan G, Garg D, Sachdev A, Matai I (2022) Biomaterials-based regenerative strategies for skin tissue wound healing. ACS Appl Bio Mater 5(5):2069–2106. https://doi.org/10.1021/acsabm.2c00035
31. Ren X, Das R, Tran P, Ngo TD, Xie YM (2018) Auxetic metamaterials and structures: a review. Smart Mater Struct. https://doi.org/10.1088/1361-665X/aaa61c
32. Liu Y, Hu H (2010) A review on auxetic structures and polymeric materials. Sci Res Essays 5(10):1052–1063
33. Shukla S, Behera BK (2022) Auxetic fibrous materials and structures in medical engineering—a review. Bioeng Transl Med. https://doi.org/10.1080/00405000.2022.2116549
34. Gupta S, Gupta V, Chanda A (2022) Biomechanical modeling of novel high expansion auxetic skin grafts. Int. J Numer Method Biomed Eng 38(5):3586. https://doi.org/10.1002/cnm.3586
35. Dudek KK, Martínez JAI, Ulliac G, Kadic M (2022) Micro-scale auxetic hierarchical mechanical metamaterials for shape morphing. Adv Mater 34(14):2110115. https://doi.org/10.1002/adma.202110115
36. Dudek KK et al (2017) On the dynamics and control of mechanical properties of hierarchical rotating rigid unit auxetics. Sci Rep 7(1):1–9. https://doi.org/10.1038/srep46529
37. Han DX, Chen SH, Zhao L, Tong X, Chan KC (2022) Architected hierarchical kirigami metallic glass with programmable stretchability. AIP Adv 12(3):8–13. https://doi.org/10.1063/5.0084906
38. Chanda A, Upchurch W (2018) Biomechanical modeling of wounded skin. J Compos Sci. https://doi.org/10.3390/jcs2040069
39. Gupta V, Chanda A (2023) Expansion potential of novel skin grafts simulants with I-shaped auxetic incisions. Biomed Eng Adv 5:100071. https://doi.org/10.1016/J.BEA.2023.100071
40. Singh G, Gupta V, Chanda A (2022) Artificial skin with varying biomechanical properties. Mater Today Proc 62:3162–3166. https://doi.org/10.1016/J.MATPR.2022.03.433
41. Gupta V, Singh G, Chanda A (2023) High expansion auxetic skin graft simulants for severe burn injury mitigation. Eur Burn J 4(1):108–120. https://doi.org/10.3390/EBJ4010011
42. Gallagher AJ, Ní Anniadh A, Bruyere K, Otténio M, Xie H, Gilchrist MD (2012) Dynamic tensile properties of human skin. International Research Council on the Biomechanics of Injury
43. Singh G, Chanda A (2023) Biofidelic gallbladder tissue surrogates. Adv Mater Process Technol. https://doi.org/10.1080/2374068X.2023.2198835
44. Singh G, Chanda A (2023) Development and biomechanical testing of artificial surrogates for vaginal tissue. Adv Mater Process Technol. https://doi.org/10.1080/2374068X.2023.2198837
45. Singh G, Chanda A (2023) Development and mechanical characterization of artificial surrogates for brain tissues. Biomed Eng Adv 5:100084. https://doi.org/10.1016/J.BEA.2023.100084
46. Gupta V, Singh G, Chanda A (2023) Development of novel hierarchical designs for skin graft simulants with high expansion potential. Biomed Phys Eng Express 9(3):035024. https://doi.org/10.1088/2057-1976/ACC661

Chapter 5
Machining Performance of Cobalt-Chromium and β-Type Titanium Biomedical Alloy

Sandeep Devgan, Amit Mahajan, and Vinod Mahajan

1 Introduction

Today, non-traditional machining approaches confirm their emergence in the fabrication sector by processing of extremely hard and intricate-shaped metals and alloys [1, 2]. Contemporary manufacturers employ a variety of sophisticated machining techniques, including electric discharge machining, water jet machining laser beam machining and electrochemical machining [3]. Among the various other established machining processes, electric discharge machining (EDM) is a fabrication method that is extremely well known. This approach of fabricating has a significant effect on the surface characteristics as well as the development of extensive subsurface layers with altered chemical composition and morphology [4]. EDM, also referred to as thermo-electric processing, involves regulated, high-energy electrical discharges which are aimed at the surface of the substrate from the tool. A particular amount of material is removed from the surface of a substrate as a consequence of the plenty of electric discharges, thereby which raise temperatures of the intended zone [5, 6]. The fabrication of various materials and alloys depends greatly on the EDM process factors such as pulse time, current, dielectric medium, voltage, type of electrode, and polarity (positive or negative).

For surface customization of AISI D2 tool steel, Guu et al. [7] contemplated using this thermoelectric technique. They stated that the toughness and the thickness of the recast layer could be improved by boosting the amount of spark discharge energy at the workpiece from the tool electrode. The current was found to be an essential variable for material removal rate (MRR) and surface roughness when EN31 tool steel was being machined by Das et al. [8] The MRR is primarily impacted by the amplification of current intensity combined with pulse-on time. An identical outcome was

S. Devgan · A. Mahajan (✉) · V. Mahajan
Department of Mechanical Engineering, Khalsa College of Engineering and Technology, Amritsar 143001, India
e-mail: amitmahajan291@gmail.com

A. Chanda et al. (eds.), *Materials for Biomedical Simulation*, Materials Horizons: From Nature to Nanomaterials, https://doi.org/10.1007/978-981-99-5064-5_5

observed by Sharif et al. [9], who identified rising current as a crucial factor affecting all of the output results of the EDMed 316L workpiece. Similarly, Mahajan et al. [10, 11] investigated the ED machining performance two different alloys such as Co–Cr and duplex stainless steel and reported that machining parameters especially current highly influenced the material removal rate of alloys. The present work devoted to the machining of cobalt-chromium and titanium using electric discharge machining. The machining performance of two biomedical alloys, i.e., Co–Cr and Ti β-type is compared, and their results are assessed in terms of material removal rate and surface integrity.

2 Materials and Methods

In order to perform the studies, rectangular blocks with the dimensions 75 mm × 40 mm × 5 mm was purchased from Metline Industries in Mumbai, India for the tests. Table 1 show the characteristics of the workpiece alloy. Utilizing the Minitab-17 statistical analysis program, the experimental layout was created using Taguchi's L9 orthogonal array. Table 2 lists the experimentally selected values for the input machining variables, including electrode, discharge current, pulse-on time, and pulse-off time. The tests were performed on a die-sinker type EDM machine (model S645 CMAX, maker OSCARMAX, Taiwan), with negative polarity settings and an identified processing time of 20 min for each run. A dielectric medium was utilized, which was deionized water.

Each experiment was run thrice on three distinct plates of alloys. The experimental L9 orthogonal array is shown in Table 3 along with their variable configurations. During the experiment, the settings for the spark gap voltage (60 V), the dielectric medium (deionized water), and the flow pressure (0.5 kgf/cm^2) remained constant. By employing an electronic weighing balance (made citizen, model CY220), via

Table 1 Properties of workpiece alloys

Properties	Units	Co–Cr	Ti-β type
Chemical composition	N.A	Cr: 28.5%, Mo: 6%, Si: 0.7%, Mn: 0.5%, Ni: 0.25%, C: 0.22%, Fe: 0.2%, P: 0.02%, Ti: 0.01% and Co: remainder	Nb: 32.74%, Zr: 7.6%, V: 1.72%, Al: 0.5%, Fe: 0.25%, Cr: 0.22%, Mo: 0.22%, Cu: 0.03%, Mn: 0.02% and Ti: remainder
Workpiece size	mm	75 × 40 × 5	
Density	g/cm^3	10	5.06
Melting range	°C	1330 °C	1573–1690
Thermal conductivity	W/mK	9.4	6.28
Specific capacity	J/kg-K	390	525

Table 2 Experimental parameters and their levels

Input parameter	Units	Level 1	Level 2	Level 3
Electrode	N.A	Graphite	Tungsten	Tungsten-Copper
Discharge current	Ampere	10	16	25
Pulse-on time	μ-seconds	60	150	200
Pulse-off time	μ-seconds	60	150	200

Table 3 Taguchi array based on the parametric combinations of input parameters

Exp. trial	Level of input variable				Actual value of input variable			
	Current	Pulse-on	Pulse-off	Electrode	Current	Pulse-on	Pulse-off	Electrode
1	1	1	1	1	10	60	60	Gr
2	1	2	2	2	10	150	150	W
3	1	3	3	3	10	200	200	W–Cu
4	2	1	2	3	16	60	150	W–Cu
5	2	2	3	1	16	150	200	Gr
6	2	3	1	2	16	200	60	W
7	3	1	3	2	25	60	200	W
8	3	2	1	3	25	150	60	W–Cu
9	3	3	2	1	25	200	150	Gr

display readings with a maximum of three decimal places, the starting and final weights were calculated for every run in order to determine the MRR. Figure 1 shows the experimental setup of the study. Furthermore, the removal of material on the processed surface was computed using Eq. 1.

$$\text{MRR}\,(\text{mm}^3/\text{min}) = \frac{1000 \times \text{mass loss of workpiece}\,(\text{g})}{\text{workpiece density}\left(\frac{\text{g}}{\text{mm}^3}\right) \times \text{machining time}} \tag{1}$$

3 Result and Discussion

The current research computes the EDM performance for the machining of two different alloys, namely cobalt-chromium and β type titanium alloy. Three attempts were carried out for both alloys and a mean value was considered to evaluate the output responses. The outcome value of each trial for Co–Cr and Ti alloys, illustrated in Table 4.

The MRR outcomes revealed that the substrate treated according to trial 8 amongst all machined titanium substrates exhibits predominant MRR (34.86 mm^3/min). Similarly, for Co–Cr alloy, also the trial 8 was considered as the highest material removal rate (23.37 mm^3/min). Therefore, the alloy specimen machined, according to trial 8,

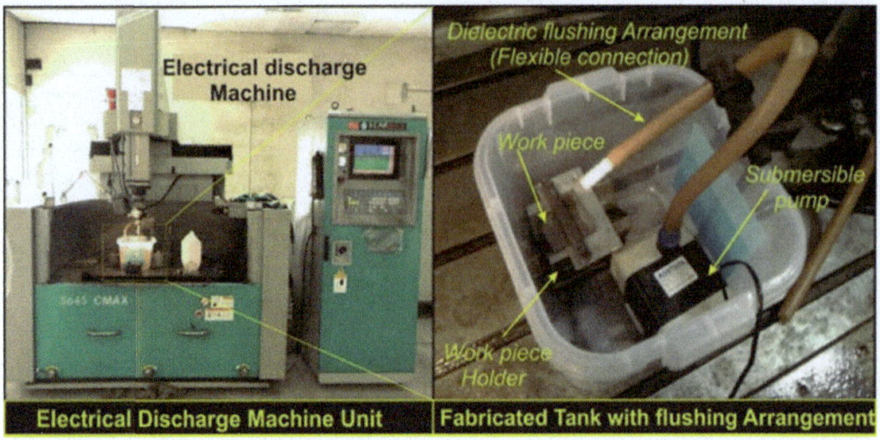

Fig. 1 Schematic view of electrical discharge machining with dielectric flushing arrangement [12]

Table 4 Output responses for EDMed Co–Cr alloy

Exp. trial	Output response value for MRR (mm^3/min.)							
	Co–Cr alloy			Mean MRR	Titanium alloy			Mean MRR
	Rep 1	Rep 2	Rep 3		Rep 1	Rep 2	Rep 3	
1	3.49	3.42	3.47	3.46	4.87	4.14	4.66	4.56
2	3.69	3.93	3.74	3.79	5.22	4.82	4.71	4.92
3	4.08	4.23	3.61	3.97	5.43	4.95	5.22	5.20
4	13.59	13.14	13.21	13.31	21.67	19.32	20.41	20.47
5	8.86	7.29	8.02	8.06	14.55	13.71	14.17	14.14
6	10.82	11.68	10.91	11.14	18.63	18.86	17.19	18.23
7	17.46	17.08	17.31	17.28	32.43	30.88	30.64	31.32
8	23.77	22.93	23.41	23.37	36.21	34.52	33.85	34.86
9	14.07	15.11	14.12	14.43	19.52	20.64	19.03	19.73

Rep Repetition

exhibits highest metal removal rate (MRR) in both alloys. At the same machining parameters, titanium alloy showed the highest metal removal rate, with an improvement of about 1.5 times greater than the Co–Cr alloy. The results also confirmed that machining at high discharge energy i.e., 16A peak current; 150 ms/60 ms pulse on/off time put significant affect on machining. It was due to the fact that the melting and evaporation of the workpiece significantly depend on discharge energy generated during the cycle. It can also be seen that the tendency of the metal removal rate for both of the alloys are same. However, the trial machined on titanium alloy showed comparatively higher highest metal removal rate except for sample 2, which was treated at low discharge energy (5 A peak current; 150 ms/150 ms pulse on/off time) among all trials. In other words, we can say that discharge energy significantly affects the material removal rate.

It can also be seen that substrates machined with Copper-tungsten (W–Cu) electrode showed comparatively higher metal removal rate than other electrodes. Thus, the Copper-tungsten (W–Cu) electrode contributed significantly to attaining the enhanced machinability. This is due to fact that copper-tungsten material has superior bulk hardening properties as compared to graphite [13]. Thus, the material removal rate was considerably higher when copper-tungsten (W–Cu) electrode was used in machining even the discharge energy was less as compared to graphite tool machined trials.

Figure 2a, b represents the surface morphology of highest material removal rate substrate (trial 8) and least material removal rate substrate (trial 1) of titanium alloy respectively. The surface roughness of highest material removal rate substrate (trial 8) was significantly higher than least material removal rate substrate (trial 1). Figure 2a represents the morphology of trial 8 surface that demonstrates pores surface at the nano-scale level ($Ra = 2.46$ μm; $Rz = 8.7$ μm). Also, the trial 8 exhibits some surface irregularities like uneven residues of molten metal and ridges of redeposit material. This is due to fact that there is a resistance in the transmission of heat and fumes due to large motel pool formed on the machining surface. This process was occured at higher peak current and large pulse on time where a large amount of discharge energies was produced. Thus, the large sized voids are shaped resulting in the formation of the non-uniform surface with higher surface irregularities [14–16]. The trial 1 surface showed the superior surface morphology ($Ra = 0.17$ μm; $Rz = 1.3$ μm) and good surface finish (Fig. 2b). Unlike, no porous structure was seen in trial 1 or at low metal removal rate substrate. Therefore, the low metal removal rate machining generates a wide range of uniformly patterned surfaces that exhibits efficient surface integrity as compare to high metal removal rate machining.

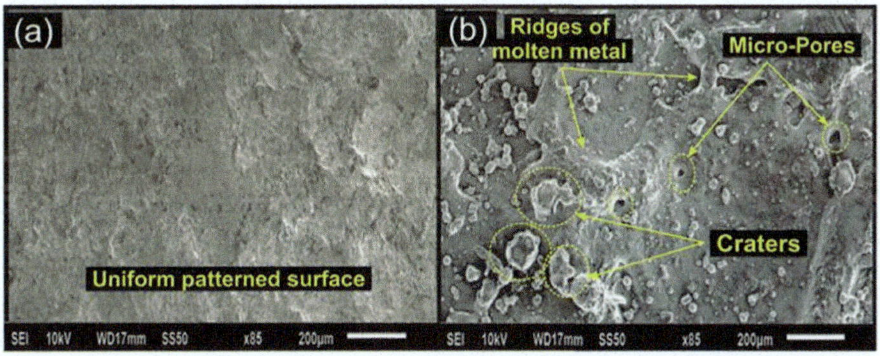

Fig. 2 FE-SEM images illustrate the surface morphology of Trial 8 and Trial 1 of titanium alloy respectively

4 Conclusions

Machining was successfully performed on the Co–Cr alloy and Ti alloy substrates by using the EDM process with the aim to figure out the machinability of alloys. The results of the material removal rate calculations confirmed that titanium alloy substrate exhibited a 1.5 times more MRR as compared with the Co–Cr alloy substrate. A 16 A peak current at a 150 ms pulse on time and a 60 ms pulse off time with a tungsten-copper (W–Cu) electrode is the best parametric set for machining. The surface morphological analysis confirmed that the low metal removal rate machining generates superior surfaces integrity as compare to high metal removal rate machining.

References

1. Kumar S, Singh R, Singh TP, Sethi BL (2009) Surface modification by electrical discharge machining: a review. J Mater Process Technol 209:3675–3687. https://doi.org/10.1016/J.JMATPROTEC.2008.09.032
2. Ho KH, Newman ST (2003) State of the art electrical discharge machining (EDM). Int J Mach Tools Manuf 43:1287–1300. https://doi.org/10.1016/S0890-6955(03)00162-7
3. Singh G, Sidhu SS, Bains PS, Bhui AS (2019) Surface evaluation of ED machined 316L stainless steel in TiO2 nano-powder mixed dielectric medium. Mater Today Proc 18:1297–1303. https://doi.org/10.1016/J.MATPR.2019.06.592
4. Mahajan A, Sidhu SS, Ablyaz T (2019) EDM Surface Treatment: An Enhanced Biocompatible Interface, Material Horizons From Nature to Nanomaterial. pp 33–40. https://doi.org/10.1007/978-981-13-9977-0_3/COVER
5. Bhui AS, Singh G, Sidhu SS, Bains PS (2018) EXPERIMENTAL INVESTIGATION OF OPTIMAL ED MACHINING PARAMETERS FOR Ti-6Al-4V BIOMATERIAL. Facta Univ Ser Mech Eng 16:337–345. https://doi.org/10.22190/FUME180824033B
6. Srivastava V, Pandey PM (2013) Study of ultrasonic assisted cryogenically cooled EDM process using sintered (Cu–TiC) tooltip. J Manuf Process 15:158–166. https://doi.org/10.1016/J.JMAPRO.2012.12.002
7. Guu YH, Hocheng H, Chou CY, Deng CS (2003) Effect of electrical discharge machining on surface characteristics and machining damage of AISI D2 tool steel. Mater Sci Eng A 358:37–43. https://doi.org/10.1016/S0921-5093(03)00272-7
8. Das MK, Kumar K, Barman TK, Sahoo P (2014) Application of artificial bee colony algorithm for optimization of mrr and surface roughness in EDM of EN31 tool steel. Procedia Mater Sci 6:741–751. https://doi.org/10.1016/J.MSPRO.2014.07.090
9. Sharif S, Safiei W, Mansor AF, Isa MHM, Saad RM (2015) Experimental study of electrical discharge machine (die sinking) on stainless steel 316L using design of experiment. Procedia Manuf 2:147–152. https://doi.org/10.1016/J.PROMFG.2015.07.026
10. Mahajan A, Devgan S, Kalyanasundaram D (2022) Surface alteration of Cobalt-Chromium and duplex stainless steel alloys for biomedical applications: a concise review. Mater Manuf Process. https://doi.org/10.1080/10426914.2022.2105873
11. Mahajan A, Singh G, Devgan S, Sidhu SS (2020) EDM performance characteristics and electrochemical corrosion analysis of Co-Cr alloy and duplex stainless steel: a comparative study. Proc Inst Mech Eng E: J Process Mech Eng 235:812–823. https://doi.org/10.1177/0954408920976739

12. Devgan S, Mahajan A, Singh G, Singh G, Sidhu SS (2022) Surface integrity of powder mixed electrical discharge treated substrate at high discharge energies. Lecture notes mechanical engineering. pp 207–217. https://doi.org/10.1007/978-981-16-2278-6_18/COVER
13. Chen YF, Chow HM, Lin YC, Lin CT (2008) Surface modification using semi-sintered electrodes on electrical discharge machining. Int J Adv Manuf Technol 36:490–500. https://doi.org/10.1007/S00170-006-0859-X
14. Singh G, Mahajan A, Devgan S, Sidhu SS (2022) Comparison of copper and tungsten electrodes for the electric discharge machined SUS-316L. Lecture notes mechanical engineering. pp 197–206. https://doi.org/10.1007/978-981-16-2278-6_17/COVER
15. Li G, Wang L, Pan W, Yang F, Jiang W, Wu X, Kong X, Dai K, Hao Y (2016) In vitro and in vivo study of additive manufactured porous Ti6Al4V scaffolds for repairing bone defects. Sci Rep 6(1):1–11. https://doi.org/10.1038/srep34072
16. Gittens RA, McLachlan T, Olivares-Navarrete R, Cai Y, Berner S, Tannenbaum R, Schwartz Z, Sandhage KH, Boyan BD (2011) The effects of combined micron-/submicron-scale surface roughness and nanoscale features on cell proliferation and differentiation. Biomaterials 32:3395–3403. https://doi.org/10.1016/J.BIOMATERIALS.2011.01.029

Chapter 6
Traction Performance of Barefoot Heel Simulant in Contaminated Bathroom Flooring Tiles

Subhodip Chatterjee, Shubham Gupta, and Arnab Chanda

1 Introduction

The bathroom region of a residential housing is considered one of the hazardous areas for slip related accidents for the older adults. Falls occurring in the bathrooms and its adjacent areas are more than twice as likely to result in a serious injury, compared to falls in other areas of the housing [1]. Movements involving motion from inside and outside of the bathtubs are also a leading cause of slip and fall accidents in older adults [1] and account for more than 70% of fall related injuries [1, 2]. Injuries leading to hospitalization, mainly from bathroom falls negatively affect the mobility, independence, and quality of life, and also induce fear, especially for older adults [3, 4]. The fear of falling, another grave consequence of falls, affects up to 85% of older adults [5] which can lead to avoiding the tendency of daily social and fundamental activities [6, 7]. A vast majority of elderly people encounter slip related accidents in the bathrooms of their respective homes. The extent of these accidents can be so severe, that it restricts the normal movement of the limbs be it the upper limbs or the lower limbs. The people who are affected by such slip related accidents are unable to bathe by themselves and require the assistance of nurses in hospitals and attendants in residences to properly perform the bathing activity. Injuries caused due to bathroom slip related accidents are associated with many adverse consequences, including increased visits to hospitals, home care services and long-term nursing home admissions [8]. Thus, evaluating the slip risk of such bathroom flooring tiles is imperative in reducing the possibility of fall related accidents.

S. Chatterjee · S. Gupta · A. Chanda (✉)
Centre for Biomedical Engineering, Indian Institute of Technology (IIT) Delhi, Delhi, India
e-mail: arnab.chanda@cbme.iitd.ac.in

A. Chanda
Department of Biomedical Engineering, All India Institute of Medical Sciences (AIIMS) Delhi, Delhi, India

A. Chanda et al. (eds.), *Materials for Biomedical Simulation*, Materials Horizons: From Nature to Nanomaterials, https://doi.org/10.1007/978-981-99-5064-5_6

Evaluation of barefoot slip risk till the present date was performed by employing human volunteers in a lab environment. In a previous study, the possibility of slips in barefoot condition in two age groups, mainly middle-aged and older adults, were quantified [9]. The friction at the foot-floor interface was measured as the volunteers manoeuvred on several dry floorings. Friction at the shoe-floor contact is quantified by dividing the forces that oppose slipping (i.e., shear forces) by the vertical force (i.e., body weight) and is typically stated as the available coefficient of friction (ACOF) [10–13]. Hence, ensuring adequate friction over common floorings is essential for the well-being of workers.

The surface roughness of the flooring was found to significantly affect the friction of the middle-aged group and was slightly higher than the older adults group, which showed minimal effect of the gender. Derler et al. [14] studied the relationship between coefficient of friction (COF) with the flooring surface roughness by performing barefoot slip simulations in wet contaminant scenarios. Fourteen individuals participated in the slip risk assessment study involving several floorings. The average ACOF measured on the floorings ranged from 0.012 to 0.50. The correlation of the floor roughness with the ACOF was found to be very low. The deviation of ACOF values was maximum on the same flooring tile was measured as 0.04, showing insignificant variability. Li et al. [15] had found out that the ACOF of different material floorings such as porcelain and ceramic floorings on the application of slippery contaminants. The floorings were substantially reported to affect the friction, while the foot slipping velocity was found to have no relation. Nagata et al. [16] investigated the effect of slip speed in barefoot condition and vertical forces in 15 older adult participants. Slip testing in a randomized way was conducted on dry, wet and soap applied flooring tiles. Siegmund et al. [17] attempted to find out the ACOF for participants moving around a shower chamber part of the bathroom consisting of the flooring tile adjacent to the bathtub in dry and water applied situations. Sixty healthy participants were selected for this study. Ten females and ten males were recruited into each of three age groups of twenty to thirty years, forty to fifty years and sixty to seventy years. The main selection criterion when recruiting volunteers for any human slip testing experiment is that the volunteers do not have any medical history of neurological disorders. Neurological disorders affect the natural gait of an individual and hence the human slip test results would not be reliable. Apart from neurological disorders, the volunteers or participants of the human slip testing experiments are also tested to make sure that they do not suffer from any physiological and musculoskeletal disorders, such as high blood pressure and muscle weakness which in turn can affect the walking pattern of the participants eventually leading to faulty data recording. Thus, the term physiologically effective implies to those participants who do not have a history of neurological, physiological and musculoskeletal disorders and whose gait cycles are normal. Force plates were placed below the bathroom flooring tile to record the vertical forces generated by the foot contact of the different participants. The friction estimated for the participants were in the range of 0.16 to 0.44. Overall barefoot friction was reported to be lower under wet conditions and in older subjects, with minimal effect of gender.

The limitations in human slip testing experiments are denoted here as the shortcomings. Various factors such as the process of getting ethical clearance for the human participants is a time-consuming process which can lead to delays in performing the slip testing experiments. The physiological and psychological parameters of the participant during a particular event of slip are not repeatable and a large number of trials as a result has to be conducted to properly understand the slip biomechanics. The human slip testing experiments are performed in a lab environment in which there are limited floorings to be tested. The variation in slip risk among larger number of floorings cannot be explored with human slip testing methodologies. The overall cost of the setup of the lab to monitor actual human slips is quite high, which is a throwback in performing greater number of human slip testing experiments. In light of these shortcomings, Nagata et al. [18] fabricated a mechanical slip risk assessment device. Barefoot COF measurement was performed by employing the flat rubber slider, across floorings with a range of surface roughness and with application of varying normal forces and sliding speeds. Barefoot COF was observed to be unaffected with variations in normal forces and sliding speeds, consistent with literature findings [9]. To date, just a handful of studies have focused on using slip testing to characterize barefoot friction. The rubber-based heel simulants not only lack in simulating the human heel's mechanical properties, but also its structural, surface, and interfacial contact properties. These shortcomings do not allow the slip testing with existing heel simulants to replace human slipping studies [19] and inhibit accurate measurement of barefoot slip risk.

In order to perform biofidelic barefoot slip testing experiments, a novel heel surrogate was developed and slip testing experiments were conducted across different bathroom flooring tiles in the presence of contaminants such as soap, shower gel, and shampoo. The second section presented the materials and methods which were employed in this study. The third section upholded the results obtained from the barefoot slip testing experiments on different floorings in the presence of different bathroom contaminants. The fourth section discussed about the conclusions obtained from the study. The primary objective of this study was to see that if the occurrence of contaminants like water, liquid soap, shower gel and shampoo had affected the barefoot slip risk probability and secondly to identify if there was any generalizable trend in the barefoot ACOF values among the different bathroom flooring tiles.

2 Materials and Methods

2.1 Design of Barefoot Surrogate

The replication of original biological appendages of human body parts is usually performed by employing a 3D scanner [20–26]. 3D scanning is also implemented to record the features of different types of footwear [27]. To replicate the slipping in barefoot condition, a laser based 3D scanner (Intel, USA) was implemented to capture

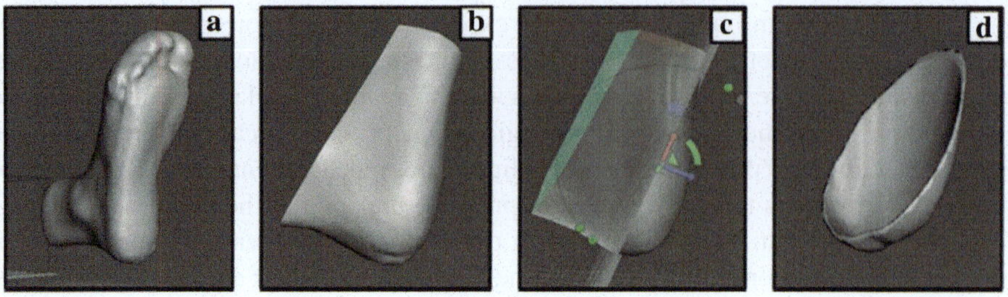

Fig. 1 Scanned 3D model of: **a** Foot, **b** Half-cut, **c** Clipping of heel geometry, **d** Heel mold

the structural attributes of the human foot. After the scanning was completed, the raw data points of the foot model were transformed into a solid model by implementing a software, namely 3D Sense (3D Systems, USA). Only the heel portion was selected to mimic the realistic slipping biomechanics as it is the initial contact zone during the event of unintentional slips [28, 29]. By using a mesh editing software, namely Meshmixer (Autodesk, USA), the heel part was extracted from the overall scanned foot. The cutting plane was inclined at an angle of 17 ± 2.5 degrees based on the average slip angle observed in the previous study [29]. Figure 1 shows the consolidated steps to prepare the heel geometry. A wall thickness of 1.5 mm was applied to generate the mold which was further used for casting the heel surrogate. 3D printing using an Ender 3 printer (Creality, China) was employed for fabricating the heel mold. An adaptor mold was further 3D printed to attach the heel surrogate with the skid tester. Polylactic Acid (PLA) filament having a diameter of 1.75 mm was employed as the material for printing the molds. The molds for 3D printed heel and adaptor are shown in Fig. 2a, b.

2.2 Fabrication of Heel Surrogate

Application of polymeric based materials such as silicone in fabricating soft tissue surrogates and orthoses is increasing [30–37]. For this study, two-part Silicone mixture (LSR 130, Chemzest, India) was mixed and fabricated to ensure similar shore hardness (i.e., 30A) as that of the plantar foot skin, in-line with the study conducted by Chanda et al. [36]. The combination mixture of Silicone was stirred for one minute and then transferred to the heel and adaptor molds (Fig. 2c). During the curing process (Fig. 2d), the partially dry molded adaptor was placed over the heel mold and was allowed to fully dry and bond adhesively with each other for 8 h. In order to perform the repeatability tests in the different slip testing situations, two heel surrogates were fabricated (Fig. 3) and slip tested on dry, water, and soap, shampoo and shower gel contaminants on eight different floorings (Table 1).

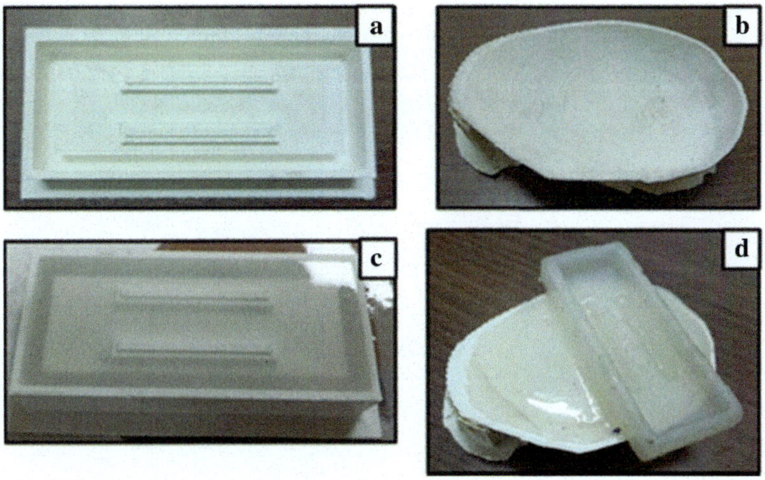

Fig. 2 **a** Attachment mold, **b** Heel mold, **c** Polymeric material placed in the mold, **d** Heel surrogate and adaptor assembly

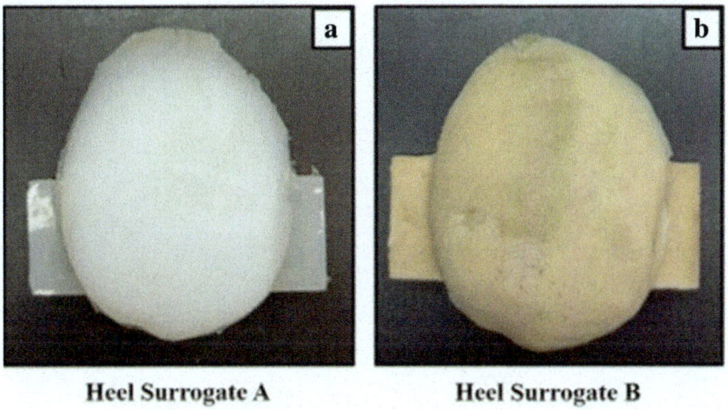

Fig. 3 **a** Surrogate A, **b** Surrogate B

2.3 Slip Testing Experiments

In order to perform human slip testing experiments, certain factors play a pivotal role. Firstly, the setup of the experiment which involves a walking platform with force plates to measure the ground force reaction at the moment of slips is required. In order to protect the participants from experiencing fall related accidents, a safety harness system is also required which will protect the participants from fall related injuries. These are the experimental considerations for human slip testing experiments. Apart from this, proper screening of the participants to ensure that they do not suffer from any neurological and musculoskeletal conditions also has to be ensured for

Table 1 Floorings and their surface roughness

Flooring designation	Flooring tile	Surface roughness (μm)
F1		69.9
F2		16.9
F3		64.3
F4		15.9
F5		18.1
F6		25.8
F7		3.4
F8		3.0

appropriate data collection during the slip testing experiments. These considerations for performing human slip testing experiments can be collectively termed as ethical considerations. Several slip risk assessment experiments involved the usage of the British pendulum skid tester. The reason for which this slip tester was used for artificial slipping experiments was due to its ease of portability, and the sliding distance of the British Pendulum Skid Tester was in accordance to that of human slips [38–42]. The heel surrogate was connected to the rectangular bottom part of the rubber slider. The connector of the heel surrogate was so prepared to enable easy attachment of the heel surrogate to the rubber slider. On attaching it to the rubber slider, the bottom portion of the heel surrogate made the required contact with the flooring tile surface. The available coefficient of friction (ACOF) was determined from the British Pendulum Number (BPN) obtained during the tests. BPN was used to calculate the ACOF by using the formula (ACOF = 0.01xBPN). ASTM E303-96 standard was used for performing the experiments using the skid tester. Prior to starting any experiments, the leveling of the tester was done by turning the leveling screws until the tester was leveled with the flooring tile. The pendulum was adjusted to such a position so that the heel surrogate just contacted the flooring tile. Five swings of the pendulum were performed on any test flooring tile surface, and the average BPN was estimated. The strike angle or the heel angle was 17 ± 2.5 degrees between the heel and the ground and was maintained during all the slip tests (Fig. 4a) based on the previous studies [29, 43, 44].

A total of eight commonly used bathroom flooring tiles were chosen for this current slip risk assessment study. Different types of tiles such as porcelain tiles, ceramic tiles, vitrified tiles, glossy tiles and super glossy tiles (Table 1) were included in the eight flooring tiles selected for study. The bathroom flooring tiles were fabricated by similar flooring tile manufacturing company known as Kajaria, India. It was found that these eight-bathroom flooring tiles were mainly being used throughout the country. The average surface roughness (Ra) of flooring tiles was focused because flooring tiles are manufactured and classified based on this parameter. The average

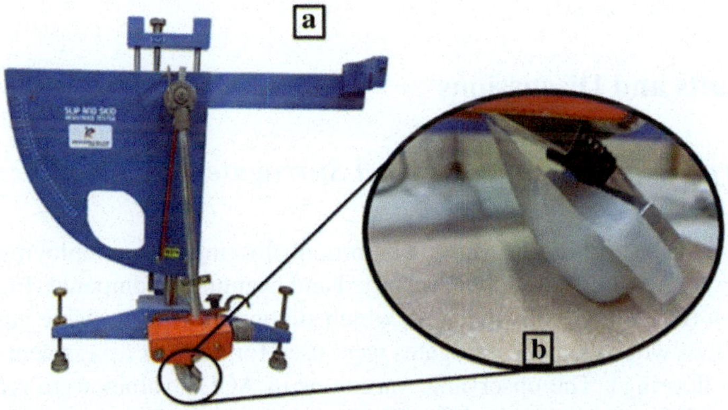

Fig. 4 **a** British pendulum slip tester, **b** Heel assembly placed under the slip tester

surface roughness (Ra) was considered in the previous research by Taylor et al. [43] in which the average surface roughness (Ra) of three different vinyl composite flooring tiles was recorded before the slip testing of 12 formal shoes. The surface roughness was evaluated using a digital surface profilometer (Sudershan Measuring Instruments, India). Apart from dry condition, contaminants such as water, liquid soap (Dettol, Reckitt, UK), shower gel (Nivea, Beiersdorf, Germany) and shampoo (Sunsilk, Unilever, UK) were tested. Considering each contaminant condition in a particular flooring, ten repeated tests were performed, and the averaged ACOF was quantified. The risk of slip was assessed by the ACOF value obtained after each slip testing experiment. If the ACOF value recorded was below 0.3, then the probability of barefoot slip increased but there is a significant possibility of recovery. In the case of a reduction of ACOF below 0.1, there is a determined chance of slipping [43, 45]. Also, the developed heel surrogates were tested for their repeatable nature on the floorings.

2.4 *Data Analysis*

Evaluation of the frictional values among the floorings in different contaminated scenarios was performed. Repeatability of the surrogates and the association between floorings were evaluated with the help of the coefficient of determination (R^2). Values more than 0.7 were designated as strong, and below 0.5 were designated to have insignificant contribution. The values varying from 0.5 to 0.7 were regarded as moderate [46, 47]. The coefficient of determination (R^2), instead of the correlation coefficient (R), was used to estimate the percentage variability of the multiple parameters on the ACOF in terms of regression. The flooring surface roughness with the varying ACOF values was quantified in dry and the other considered contaminated floorings. Also, the association of ACOF was explored across flooring conditions to establish the generalizability of barefoot slip risk.

3 Results and Discussions

3.1 *Repeatability Nature of Heel Surrogate*

After dry slip testing on the various bathroom flooring tiles, employing two heel surrogates, the ACOF values were recorded and eventually compared (Fig. 5). High correlation ($R^2 = 0.81$) was reported, which supported the repeatable nature of the ACOF values when the heel surrogates were used for slip risk assessment on similar bathroom floorings. The observable variations in ACOF values were recorded for tiles F3 and F4, which amounted to 0.05. Slightly lower variations among the ACOF

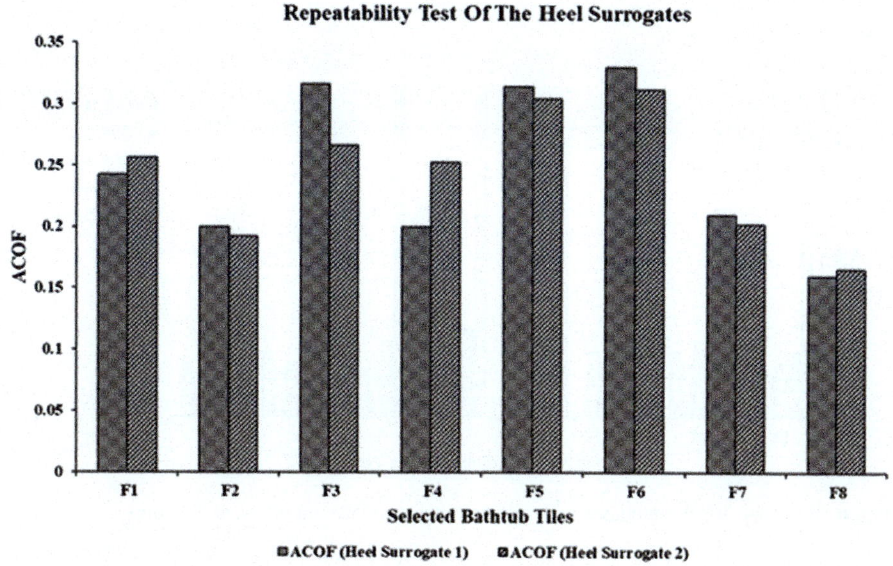

Fig. 5 Repeatable nature of the heel surrogates

values were noticed for F1 and F6 which was reported to be 0.01. The lowest differences in the ACOF values were recorded for the tiles F2, F5, F7 and F8 which amounted to 0.01. It was also seen that for ACOF values above 0.25 exhibited high difference in the ACOF values across heel surrogates. The similarity in the traction performance of the two heel surrogates which is in terms of the similar ACOF values recorded in each flooring tile indicates the quantitative comparison between the ACOF values recorded. The differences between the ACOF values of the two developed heel surrogates are low and are found to support the repeatable nature of the heel surrogate [37].

3.2 Barefoot ACOF Outcomes with Different Contaminants on Different Floorings

It was observed that the ACOF values recorded during dry slipping were the highest among all the different slip testing scenarios in this present study (Fig. 6). The variation of the ACOF values in the dry slip testing ranged from 0.17 to 0.31. The flooring tile designated as F6 showed the greatest ACOF value (i.e., 0.31) and the flooring tile designated as F8 showed the lowest ACOF value (i.e., 0.17). In the condition for wet slipping, there was a decrease in the ACOF values as compared to that of dry slip testing. The variation of the ACOF values in the wet slip testing ranged from 0.05 to 0.10. It was observed in this wet slip testing scenario, the ACOF value

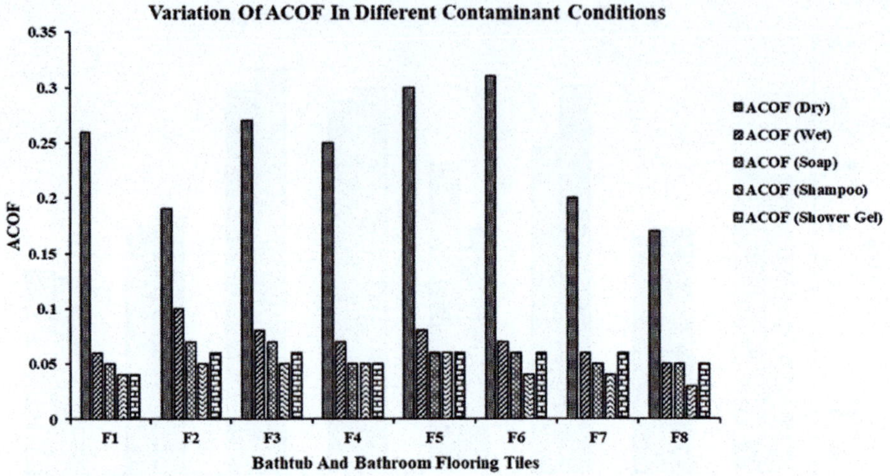

Fig. 6 Barefoot ACOF outcomes with different contaminants on different floorings

(i.e., 0.10) exhibited by the tile F2 was the highest and the flooring tile designated as F8 exhibited the lowest ACOF value (i.e., 0.05). There was a further decrease in the ACOF values when soap was employed as the contaminant. The range of ACOF values was from 0.05 to 0.07. Similar to that of the wet slip testing scenario, the flooring tile designated as F2 showed the greatest ACOF value (i.e., 0.07) and the flooring tile designated as F8 showed the lowest ACOF value (i.e., 0.05). The ACOF values for the soap and shower gel contaminants were in a similar range (i.e., 0.05 to 0.06). The lowest ACOF values were recorded for the shampoo contamination, varying from 0.03 and went upto 0.06. Across the floorings, the tile designated as F5 displayed the highest ACOF (i.e., 0.06) and the tile F8 exhibited the lowest ACOF (i.e., 0.03) in shampoo contaminated condition. The main information obtained from this graph is that how the traction barefoot performance decreased when transitioning from dry slip testing to wet and subsequently bathroom contaminant applied slip testing conditions. The slip probability increased when transitioning from wet to soap, shampoo and shower gel applied condition as evident from the decrease in ACOF values. Similarity in the traction performance in shampoo and shower gel applied conditions were observed in the F1, F4 and F5 bathroom flooring tiles.

3.3 Variation of Barefoot ACOF with Surface Roughness

An important observation was made when dry slipping was performed on the bathroom flooring tiles specifically F1, F2, F3, F4 and F8 showed a direct trend among the surface roughness and the ACOF (Fig. 7a). The bathroom flooring tiles F5 and F6 had almost similar and high ACOF values, but with wide variation in the surface roughness values. Considering the slip testing in wet condition (Fig. 7b), a decrease in the barefoot traction was observed. Some particular bathroom flooring tiles such as F3, F4, F7 and F8 showed a positive correlation among the barefoot ACOF, and the surface roughness values. The bathroom flooring tiles F1, F7, F4, and F6 had similar ACOF values but different surface roughness values. Specifically, two bathroom flooring tiles F3 and F5 had similar high ACOF values but there was a substantial difference in the surface roughness between them. The number of flooring tiles exhibiting similar ACOF values increased when soap was adopted as the contaminant (Fig. 7c). The flooring tiles F1, F4, F7 and F8 showed similar ACOF values but widely different surface roughness values. The highest ACOF values obtained in this case were that of the flooring tiles F2 and F3, which had a wide variation in their surface roughness values. The lowest obtained ACOF values were from slip testing with the shampoo (Fig. 7d). The flooring tiles F1, F6 and F7, and similarly the tiles F2, F3 and F4 exhibited similar ACOF values but there were wide differences in the surface roughness values. There was a similarity between the ACOF variation trend with the shower gel (Fig. 7e) and the liquid soap applied condition. The flooring tiles F2, F3, F5, F6 and F7 exhibited similar ACOF values but with different surface roughness values.

3.4 Correlation Between Barefoot ACOF and Surface Roughness

It was observed that the overall correlation among the ACOF and surface roughness was poor, an attempt was made to isolate at least five floorings for each contaminated condition, for which meaningful correlations existed. The main focus of this part of the study was to find the relation between the surface roughness of the bathroom flooring tiles and the traction performance recorded by the heel surrogate on these flooring tiles. Initially, the analysis was performed for all the eight flooring tiles but an attempt was made to identify the results of those floorings which had high correlation between the surface roughness and the ACOF values. On the basis of this classification, minimum 5 floorings were considered as the threshold for the correlation analysis for which the term "isolating at least five floorings" were used. This exercise was specifically conducted to study the possibility of barefoot ACOF prediction from known surface roughness of floorings, in the presence of different contaminants. During the process of forming these different flooring combinations, two floorings which negatively affected the correlation were removed. In dry slip

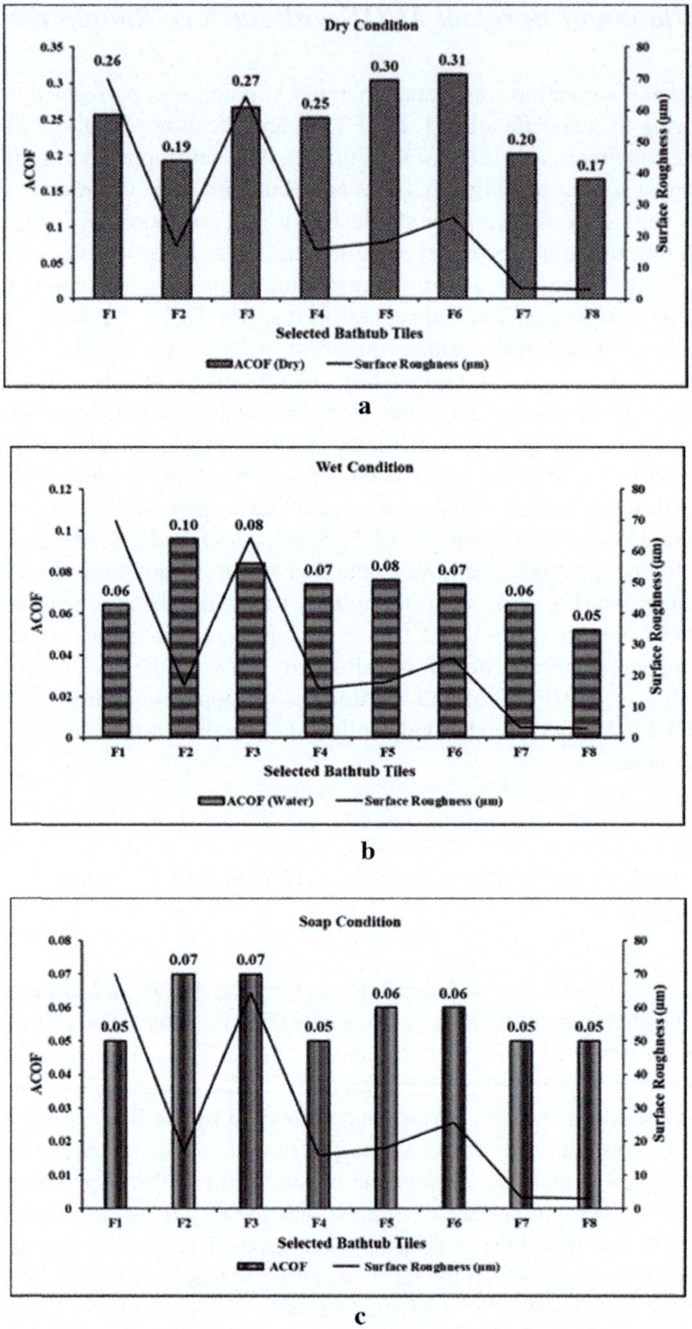

Fig. 7 Variation of barefoot ACOF with surface roughness in: **a** Dry, **b** Wet, and with contaminants: **c** Soap, **d** Shampoo, and **e** Shower gel

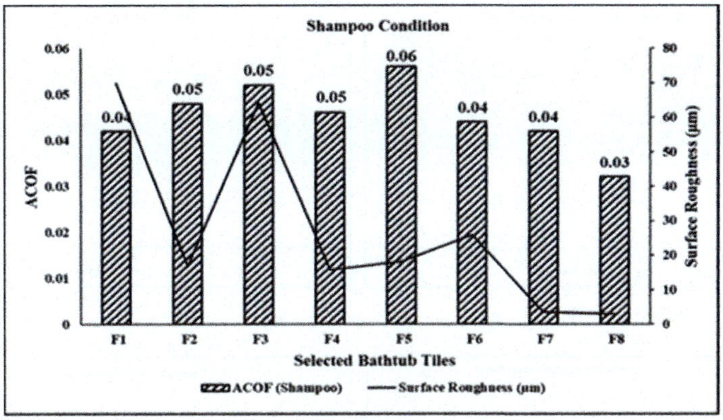

d

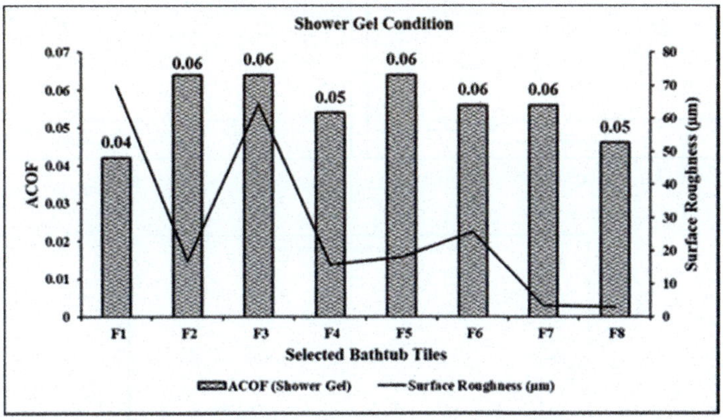

e

Fig. 7 (continued)

testing (Fig. 8a), the flooring tiles F1, F2, F3, F4, F7 and F8 displayed moderate correlation ($R^2 = 0.62$) between the surface roughness values and the ACOF values. In the wet condition (Fig. 8b), the bathroom flooring tiles F3, F4, F5, F6, F7 and F8 showed moderate correlation ($R^2 = 0.65$) among the ACOF and the surface roughness values. For slip testing with the liquid soap (Fig. 8c), the same set of flooring tiles exhibited moderate correlation ($R^2 = 0.57$) amongst the ACOF and the surface roughness. In case of the shampoo applied condition (Fig. 8d), the flooring tiles F2, F3, F4, F7 and F8 showed moderate correlation ($R^2 = 0.55$) among the ACOF and the surface roughness. For the shower gel applied condition (Fig. 8e), the flooring tiles F1, F2, F4, F5, F6, and F7 showed a reasonable correlation ($R^2 = 0.61$) between the ACOF and the surface roughness.

Surface roughness was not found to be a determining factor in estimating the ACOF values on the bathroom flooring tiles when liquid soap, shower gel and

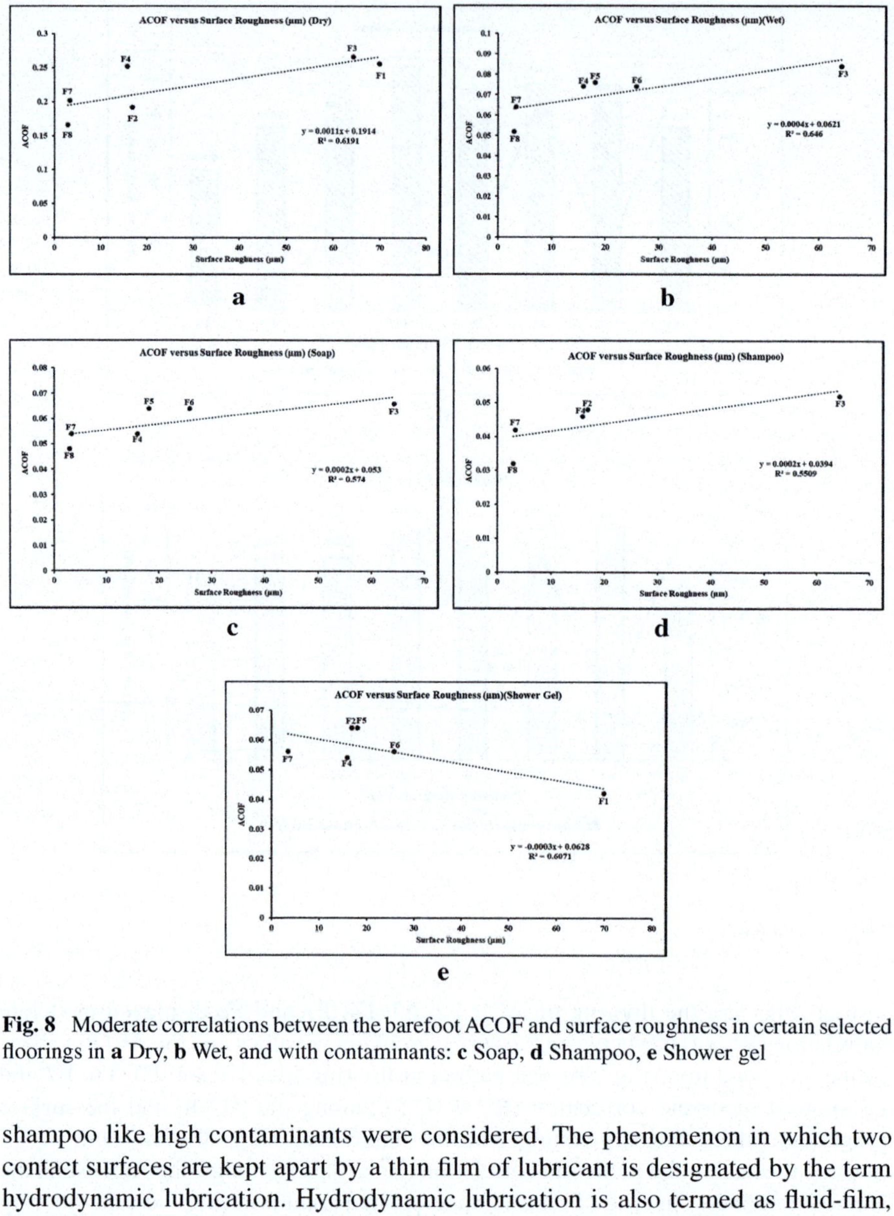

Fig. 8 Moderate correlations between the barefoot ACOF and surface roughness in certain selected floorings in **a** Dry, **b** Wet, and with contaminants: **c** Soap, **d** Shampoo, **e** Shower gel

shampoo like high contaminants were considered. The phenomenon in which two contact surfaces are kept apart by a thin film of lubricant is designated by the term hydrodynamic lubrication. Hydrodynamic lubrication is also termed as fluid-film, thick-film or flooded lubrication. A film of lubricant is accumulated between the surfaces of the contacting bodies in relative motion. It was found that with the same load, the pressure developed in the film increases as the viscosity of the fluid increases [48]. The film formation might have reduced the effect of the valleys and ridges present on the topography of the flooring tiles. The surface film formation occurs due to the presence of slippery contaminants such as water, liquid soap, shower gel and shampoo resulting in lower correlation values between the barefoot ACOF and surface roughness. Reduction of the effect of surface roughness occurs due to the formation of the surface film and with increasing viscosity of the contaminant, the correlation between the barefoot ACOF and surface roughness decreases. The high viscous fluid film diminishes the effect of the surface roughness of the bathroom flooring tiles used. The surface irregularities of the flooring tiles are completely submerged in the viscous fluid film. The moment when the heel surrogate comes in contact with the flooring tile, the viscous fluid film prevents the actual heel strike contact. Thus, the ACOF decreases with increasing ACOF, as the surface roughness effect is completely diminished by the presence of the viscous shower gel contaminant.

3.5 Generalizable Barefoot ACOF Across Floorings and Contaminants

The traction performance of the heel surrogate varied across the different bathroom flooring tiles in the presence of different bathroom contaminants. It was observed that on comparing the trend in the variation of the traction performance of the heel surrogate among the different bathroom flooring tiles, a certain degree of similarity in the variation of traction performance was observed. This similarity in the variation of the barefoot traction performance among the different bathroom flooring tiles is referred to as the generalizability of barefoot slip risk. The generalizability of barefoot slip risk among the eight different bathroom flooring tiles was observed to increase as the viscosity of the contaminant increased. Thus, the generalizable trend increased from transitioning from water to shower gel contaminant. Slip testing was performed across different flooring tiles in the presence of different contaminants and the correlation analysis among the barefoot ACOF values for the different flooring tiles considering each contaminant scenario was performed accordingly. In the case of the dry slip testing (Fig. 9a), one flooring combination F4-F7 exhibited moderate correlation ($0.5 < R^2 < 0.7$) and five flooring groups F2-F3, F1-F5, F1-F6, F2-F7 and F3-F7 showed high correlations ($R^2 > 0.7$). In wet condition (Fig. 9b), majority of the combinations showed low correlation ($R^2 < 0.7$) and just three flooring groups F1-F3, F1-F4, and F3-F4 showed high correlation. With soap as the contaminant, all the flooring combinations were found to have low correlations (Fig. 9c). With shampoo as the contaminant (Fig. 9d), five flooring combinations F2-F3, F5-F6, F2-F7, F3-F7 and F1-F8 exhibited high correlations. Six flooring combinations F1-F2, F5-F6, F5-F7, F6-F7, F1-F8, and F2-F8 exhibited high correlations in the shower gel contaminated condition (Fig. 9e). Overall, the most generalizable flooring combinations were identified from barefoot slip testing with shower gel contamination, followed by shampoo, dry, and wet conditions. No generalizability was observed

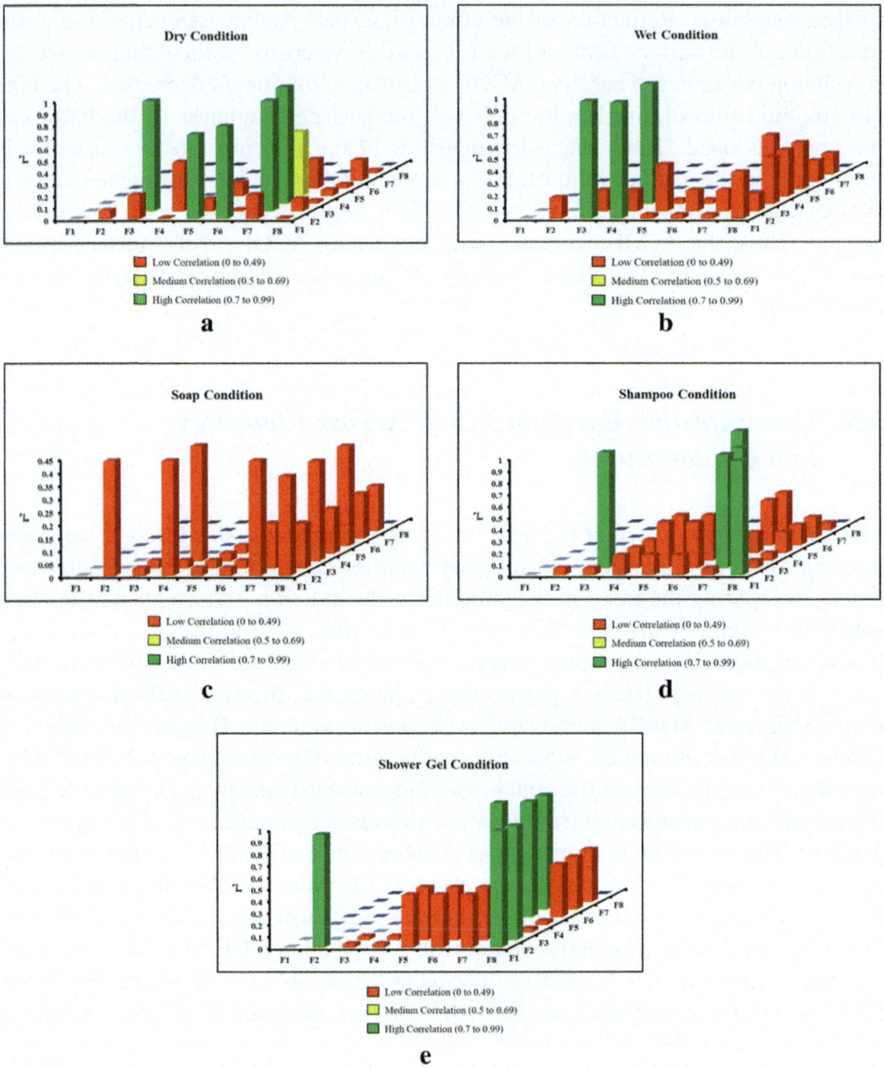

Fig. 9 Correlation among the flooring tiles in: **a** Dry, **b** Wet, and with contaminants: **c** Soap, **d** Shampoo, **e** Shower gel

across any floorings for the most common soap contaminated condition. These findings indicate the need to test only a few floorings for barefoot slip risk assessment in dry and majority of the slippery conditions, except in the case of soap, where all floorings need to be tested at least once to estimate the slip risk.

4 Conclusions

In this work, the barefoot slip risk quantification was performed by utilizing a biofidelic and as well as a tribofidelic human heel surrogate, which was fabricated by the process of 3D scanning and 3D printing. Eight bathroom flooring tiles were employed for slip risk assessment study in certain contaminant conditions which included dry, wet, soap, shower gel, and shampoo conditions. The friction in the barefoot condition was observed to substantially decrease on transitioning from dry to contaminated conditions. The floorings exhibiting meaningful correlations were estimated for all the contaminants. There were some particular flooring tiles specifically in dry, wet, shampoo and shower gel contaminant conditions which exhibited similar trends in traction performance. Slip risk assessment in these flooring tiles yielded similar trend in variation of the ACOF values also. These observations clarify the barefoot slip risk probability in different bathroom flooring tiles in the presence of commonly availed bathroom contaminants. The main focus of this present study was to understand the barefoot traction performance in dry and viscous contaminant scenarios for bathroom flooring tiles. Future studies will focus on a larger number of bathroom flooring tiles so that more detailed statistical analysis can be performed.

Acknowledgements No funding was received for this work.

Statements and Declarations The authors declare no conflict of interest with respect to the research, authorship, and/or publication of this article.

References

1. Stevens JA, Haas EN, Haileyesus T (2011) Nonfatal bathroom injuries among persons aged ≥15 years-United States, 2008. J Safety Res 42:311–315. https://doi.org/10.1016/j.jsr.2011.07.001
2. Yiannakoulias N, Rowe BH, Svenson LW, Schopflocher DP, Kelly K, Voaklander DC et al (2008) The epidemiology of bathing disability in older persons. J Am Geriatr Soc 5:311–315. https://doi.org/10.5249/jivr.v5i1.177
3. Scheffer AC, Schuurmans MJ, Van dijk N, Van der hooft T, De rooij SE (2008) Fear of falling: measurement strategy, prevalence, risk factors and consequences among older persons. Age Ageing 37:19–24. https://doi.org/10.1093/ageing/afm169.
4. Vellas BJ, Wayne SJ, Romero LJ, Baumgartner RN, Garry PJ (1997) Fear of falling and restriction of mobility in elderly fallers. Age Ageing 26:189–193. https://doi.org/10.1093/ageing/26.3.189
5. Howland J, Lachman ME, Peterson EW, Cote J, Kasten L, Jette A (1998) Covariates of fear of falling and associated activity curtailment. Gerontologist 38:549–555. https://doi.org/10.1093/geront/38.5.549
6. Deshpande N, Metter EJ, Lauretani F, Bandinelli S, Guralnik J, Ferrucci L (2008) Activity restriction induced by fear of falling and objective and subjective measures of physical function: a prospective cohort study. J Am Geriatr Soc 56:615–620. https://doi.org/10.1111/j.1532-5415.2007.01639.x

7. Brouwer B, Musselman K, Culham E (2004) Physical function and health status among seniors with and without a fear of falling. Gerontology 50:135–141. https://doi.org/10.1159/000076771
8. Brody KK (1997) Evaluation of a self-report screening instrument to predict frailty outcomes in aging populations. Gerontologist 37:182–191. https://doi.org/10.1093/geront/37.2.182
9. Rozin Kleiner AF, Galli M, Araujo do Carmo A, Barros RML (2015) Effects of flooring on required coefficient of friction: Elderly adult vs. middle-aged adult barefoot gait. Appl Ergon 50:147–52. https://doi.org/10.1016/j.apergo.2015.02.010.
10. Hemler SL, Pliner EM, Redfern MS, Haight JM, Beschorner KE (2020) Traction performance across the life of slip-resistant footwear: preliminary results from a longitudinal study. J Safety Res 74:219–225. https://doi.org/10.1016/J.JSR.2020.06.005
11. Gupta S, Sidhu SS, Chatterjee S, Malviya A, Singh G, Chanda A (2022) Effect of floor coatings on slip-resistance of safety shoes. Coatings 12:1455. https://doi.org/10.3390/COATINGS1210 1455
12. Gupta S, Chatterjee S, Malviya A, Chanda A (2022) Traction performance of common formal footwear on slippery surfaces. Surfaces 5:489–503 2022. https://doi.org/10.3390/SURFACES5 040035.
13. Gupta S, Malviya A, Chatterjee S, Chanda A (2022) Development of a portable device for surface traction characterization at the shoe–floor interface. Surfaces 5:504–520. https://doi.org/10.3390/SURFACES5040036
14. Derler S, Huber R, Feuz HP, Hadad M (2009) Influence of surface microstructure on the sliding friction of plantar skin against hard substrates. Wear 267:1281–1288. https://doi.org/10.1016/j.wear.2008.12.053
15. Li KW, Wen HC (2014) Friction between foot and floor under barefoot conditions: a pilot study. IEEE Int Conf Ind Eng Eng Manag 1651–5. https://doi.org/10.1109/IEEM.2013.6962690
16. Nagata H, Kato M, Watanabe H, Inoue Y, Kim IJ (2008) A preliminary study on slip potentials of stepping barefoot on slippery floors. Contemp Ergon 2008:710–716
17. Siegmund GP, Flynn J, Mang DW, Chimich DD, Gardiner JC (2010) Utilized friction when entering and exiting a dry and wet bathtub. Gait Posture 31:473–478. https://doi.org/10.1016/j.gaitpost.2010.02.003
18. Nagata H, Watanabe H, Inoue Y, Kim I (2008) Development of a slip-resistance meter for evaluating fall risk on slippery floors covered with soapsuds
19. Iraqi A, Cham R, Redfern MS, Vidic NS, Beschorner KE (2018) Kinematics and kinetics of the shoe during human slips. J Biomech 74:57–63. https://doi.org/10.1016/J.JBIOMECH.2018.04.018
20. Wan FKW, Yick KL, Yu WWM (2017) Validation of a 3D foot scanning system for evaluation of forefoot shape with elevated heels. Meas J Int Meas Confed 99:134–144. https://doi.org/10.1016/j.measurement.2016.12.005
21. Yamashita T, Yamashita K, Sato M, Kawasumi M, Ata S (2021) Foot-surface-structure analysis using a smartphone-based 3D foot scanner. Med Eng Phys 95:90–96. https://doi.org/10.1016/j.medengphy.2021.08.001
22. Chen LH, Chang CC, Wang MJ, Tsao L (2018) Comparison of foot shape between recreational sprinters and non-habitual exercisers using 3D scanning data. Int J Ind Ergon 68:337–343. https://doi.org/10.1016/j.ergon.2018.08.006
23. Chanda A, Unnikrishnan V (2018) Novel insole design for diabetic foot ulcer management. Proc Inst Mech Eng Part H J Eng Med 232:1182–1195. https://doi.org/10.1177/095441191880 8330
24. Singh G, Gupta S, Chanda A (2021) Biomechanical modelling of diabetic foot ulcers: a computational study. J Biomech 127:110699. https://doi.org/10.1016/J.JBIOMECH.2021.110699
25. Gupta S, Singh G, Chanda A (2021) Prediction of diabetic foot ulcer progression: a computational study. Biomed Phys Eng Express 7:065020. https://doi.org/10.1088/2057-1976/AC29F3
26. Irzmańska E, Okrasa M (2018) Evaluation of protective footwear fit for older workers (60+): a case study using 3D scanning technique. Int J Ind Ergon 67:27–31. https://doi.org/10.1016/j.ergon.2018.04.001

27. McGorry RW, DiDomenico A, Chang CC (2010) The anatomy of a slip: kinetic and kinematic characteristics of slip and non-slip matched trials. Appl Ergon 41:41–46. https://doi.org/10.1016/j.apergo.2009.04.002
28. Grönqvist R, Hirvonen M, Rajamäki E, Matz S (2003) The validity and reliability of a portable slip meter for determining floor slipperiness during simulated heel strike. Accid Anal Prev 35:211–225. https://doi.org/10.1016/S0001-4575(01)00105-1
29. Chanda A, Jones TG, Beschorner KE (2018) Generalizability of footwear traction performance across flooring and contaminant conditions. IISE Trans Occup Ergon Hum Factors 6:98–108. https://doi.org/10.1080/24725838.2018.1517702
30. Singh G, Chanda A (2021) Mechanical properties of whole-body soft human tissues: a review. Biomed Mater 16:062004. https://doi.org/10.1088/1748-605X/AC2B7A
31. Singh G, Gupta V, Chanda A (2022) Mechanical characterization of rotating triangle shaped auxetic skin graft simulants. Facta Univ Ser Mech Eng. https://doi.org/10.22190/FUME220226038S
32. Singh G, Gupta V, Chanda A (2022) Artificial skin with varying biomechanical properties. Mater Today Proc. https://doi.org/10.1016/J.MATPR.2022.03.433
33. Makode S, Singh G, Chanda A (2021) Development of novel anisotropic skin simulants. Phys Scr 96:125019. https://doi.org/10.1088/1402-4896/AC2EFD
34. Chanda A, Chatterjee S, Gupta V (2020) Soft composite based hyperelastic model for anisotropic tissue characterization. J Compos Mater 54:4525–4534. https://doi.org/10.1177/0021998320935560
35. Chanda A, Callaway C, Clifton C, Unnikrishnan V (2018) Biofidelic human brain tissue surrogates. Mech Adv Mater Struct 25:1335–1341. https://doi.org/10.1080/15376494.2016.1143749
36. Chanda A, McClain S (2019) Mechanical modeling of healthy and diseased calcaneal fat pad surrogates. Biomimetics 4:1. https://doi.org/10.3390/biomimetics4010001
37. Chatterjee S, Chanda A (2022) Development of a tribofidelic human heel surrogate for barefoot slip testing. J Bionic Eng. https://doi.org/10.1007/s42235-021-00138-0
38. Nagata H, Watanabe H, Inoue Y, Kim I (2016) Fall risks and validities of various methods to measure frictional properties of slippery floors covered with soapsuds θ.
39. Terjék A, Dudás A (2018) Ceramic floor slipperiness classification—a new approach for assessing slip resistance of ceramic tiles. Constr Build Mater 164:809–819. https://doi.org/10.1016/j.conbuildmat.2017.12.242
40. Sudoł E, Szewczak E, Małek M (2021) Comparative analysis of slip resistance test methods for granite floors. Materials (Basel) 14:1–15. https://doi.org/10.3390/ma14051108
41. Gupta S, Chatterjee S, Chanda A (2022) Effect of footwear material wear on slips and falls. Mater Today Proc. https://doi.org/10.1016/J.MATPR.2022.04.313
42. Chatterjee S, Gupta S, Chanda A (2022) Barefoot slip risk assessment of Indian manufactured ceramic flooring tiles. Mater Today Proc. https://doi.org/10.1016/J.MATPR.2022.04.428
43. Jones T, Iraqi A, Beschorner K (2018) Performance testing of work shoes labeled as slip resistant. Appl Ergon 68:304–312. https://doi.org/10.1016/j.apergo.2017.12.008
44. Iraqi A, Vidic NS, Redfern MS, Beschorner KE (2020) Prediction of coefficient of friction based on footwear outsole features. Appl Ergon 82:102963. https://doi.org/10.1016/j.apergo.2019.102963
45. Hemler SL, Charbonneau DN, Iraqi A, Redfern MS, Haight JM, Moyer BE et al (2019) Changes in under-shoe traction and fluid drainage for progressively worn shoe tread. Appl Ergon 80:35–42. https://doi.org/10.1016/J.APERGO.2019.04.014
46. Chanda A, Jones TG, Beschorner KE (2018) Generalizability of footwear traction performance across flooring and contaminant conditions. 6:98–108. https://doi.org/10.1080/24725838.2018.1517702
47. Chanda A, Reuter A, Beschorner KE (2019) Vinyl composite tile surrogate for mechanical slip testing. 7:132–41. https://doi.org/10.1080/2472583820191637381
48. Gohar R, Safa MMA (2010) Fluid film lubrication. Tribol Dyn Engine Powertrain Fundam Appl Futur Trends 132–70. https://doi.org/10.1533/9781845699932.1.132

Chapter 7
Development and Biomechanical Testing of Human Stomach Tissue Surrogates

Gurpreet Singh and Arnab Chanda

1 Introduction

The stomach tissue is a muscular sac near the abdomen, located between the duodenum and esophagus. It is the major functional organ of the gastrointestinal system, regulates the absorption timing for large meals, and acts as a nutrient reservoir. The gastric enzymes and fluids transform partly digested food into a semi-liquid substance [1]. Even after decades of research on the biomechanics of soft tissues, the mechanical properties of the human stomach have been studied by few researchers. The geometry of stomach tissue is more intricate than that of other tissues of the gastrointestinal system, which makes the mechanical properties of the human stomach tissue more important [2]. Most of the available literature has been reported on animal models [3–6], and few studies have investigated the biomechanical properties of the cadaveric stomach tissue [7, 8]. Egorov et al. [7] studied the mechanical properties of the cadaver stomach under axial and transverse tensile loading conditions. Lim et al. [8] used the cadaveric stomach tissue sample under mechanical loading and evaluated the elastic modulus. Ethical and biosafety concerns with the handling and testing of cadaveric samples were the primary reason for the limited studies on human stomach tissue. The mechanical behavior of the animal models cannot be translated to the results of mechanical behavior using the cadaveric tissues. The knowledge of the mechanical properties of the hollow gastrointestinal functional tissues (e.g., stomach tissue) is indispensable for a better understanding of their physiology, biomechanics, and ability to withstand distension [7]. The human abdominal functional tissues (e.g., stomach tissue) are often prone to traumatic injuries

G. Singh · A. Chanda (✉)
Centre for Biomedical Engineering, Indian Institute of Technology (IIT) Delhi, Delhi, India
e-mail: arnab.chanda@cbme.iitd.ac.in

A. Chanda
Department of Biomedical Engineering, All India Institute of Medical Sciences (AIIMS) Delhi, Delhi, India

A. Chanda et al. (eds.), *Materials for Biomedical Simulation*, Materials Horizons: From Nature to Nanomaterials, https://doi.org/10.1007/978-981-99-5064-5_7

due to automotive crashes, athletic activities, explosions, and other dynamic impact situations.

In this work, a four-part silicone-based polymeric material system was designed to fabricate the candidate samples for stomach tissue surrogates and tested under uniaxial tensile loading conditions. The polymeric material of varying shore hardness was arbitrarily mixed, and 15 candidate samples for stomach tissue surrogates were fabricated. Mooney–Rivlin and Yeoh hyperelastic formulations were used to evaluate the non-linear behavior of the developed tissue surrogates. The prediction accuracies of both models were determined using the R^2 correlation index. The developed tissue surrogates with realistic mechanical properties are anticipated to provide a plethora of applications, including trauma research, surgical training, diagnosis purposes, and understanding the biomechanics of the stomach tissue for different injury models. In addition, the developed tissue surrogates are easy to handle, castable in any shape, and have no ethical and biosafety issues like the cadaveric tissues. Section 2 elaborates on the fabrication, mechanical testing, and material modeling of the candidate samples for stomach tissue surrogates. The results of the tested samples, repeatability tests of the controlled samples, and coefficients of hyperelastic curve fit models are discussed in Sect. 3, followed by conclusions of the work in Sect. 4.

2 Materials and Methods

2.1 3D-Printing of Designed Mold

SolidWorks (Dassault Systèmes, Vélizy-Villacoublay, France), a computer-aided design (CAD) software, was used to design a mold for fabricating test coupons of identical dimensions (5 cm × 1 cm × 3 mm). The designed mold was saved in stereolithographic (STL) format and 3D-printed on a Voxelab Aquila (Zhejiang Flashforge 3D Technology Co., Ltd., China) using Polylactic acid (PLA) material. Figure 1 shows the SolidWorks design and the 3D-printed mold used for fabricating the test coupons.

2.2 Fabrication of Candidate Samples for Tissue Surrogates

The candidate samples for stomach tissue surrogates were fabricated with a four-part material system. The two-part polymeric material of one shore hardness was mixed with another two-part polymeric material to obtain the four-part mixture. The selected polymeric materials were of three shore hardness, i.e., 5A, 15A, 30A (Chemzest Enterprises, Chennai, India), and arbitrarily mixed in the specified ratio to fabricate a total of 15 test coupons (see Table 1). The shore hardness of the two-part silicone-based polymeric material was measured as per the ASTM D2240 standard

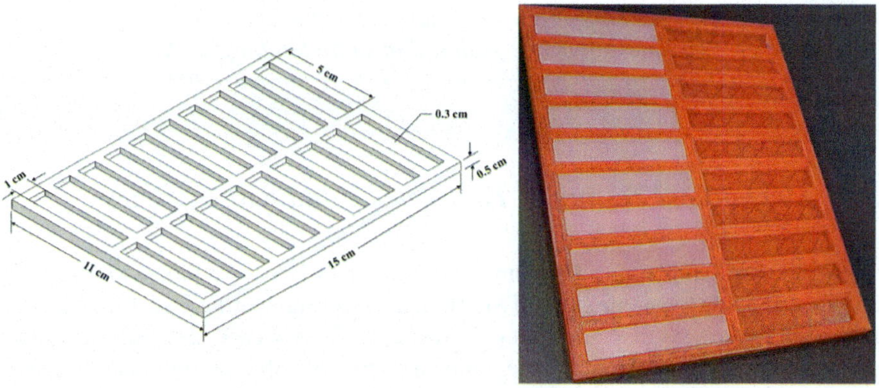

(a) (b)

Fig. 1 **a** SolidWorks designed mold, **b** 3D-printed mold with fabricated candidate samples

Table 1 Material composition for candidate stomach tissue surrogates (by wt. %)

Test coupon	Material 1		Material 2		Material 3		Shore hardness
	Part A	Part B	Part A	Part B	Part A	Part B	
1	50	50	–	–	–	–	5A
2	45	45	5	5	–	–	6A
3	35	35	15	15	–	–	8A
4	25	25	25	25	–	–	10A
5	15	15	35	35	–	–	12A
6	5	5	45	45	–	–	14A
7	–	–	50	50	–	–	15A
8	–	–	47	47	3	3	16A
9	–	–	40	40	10	10	18A
10	–	–	33	33	17	17	20A
11	–	–	27	27	23	23	22A
12	–	–	20	20	30	30	24A
13	–	–	13	13	37	37	26A
14	–	–	7	7	43	43	28A
15	–	–	–	–	50	50	30A

using the Shore durometer scale [9]. The test coupons of identical dimensions (50 mm in length, 10 mm in width, and 3 mm in thickness) were prepared by varying the proportion of four-part polymeric material (by weight %). The four-part silicone-based polymeric material was thoroughly mixed and left to cure for 5–6 h. Table 1 shows the compositional detail of the 15 unique test coupons used in the present work.

Material 1: Two-part polymeric material of shore hardness 5A
Material 2: Two-part polymeric material of shore hardness 15A
Material 3: Two-part polymeric material of shore hardness 30A

2.3 Mechanical Testing of Fabricated Candidate Samples

The candidate samples for the stomach tissue surrogates were uniaxially tested on a universal testing machining (UTM) (Finetechno Engineering Pvt. Ltd., Kolkata, India), and the force-displacement data was recorded for each test. Some considerations should be considered for the uniaxial testing of soft materials such as human soft tissues and polymeric materials [2, 10–13]. First, the UTM tensile test on soft materials requires special grips that provide sufficient friction to prevent slippage. Second, soft materials should be tested at a specific strain rate to compare the results and findings with the literature. Third, the responses to the tensile testing results may vary due to the shape and size of the sample [10, 13, 14]. All three parameters were considered in the present study, and the experimental framework was designed accordingly. The candidate samples for stomach tissue surrogates were properly gripped, and a specific sample size was used to avoid slipping during the experiments. Each specimen was tested at six different strain rates: 0.16, 0.4, 0.5, 0.8, 1, and 2.5 mm/s. These strain rates were determined from the literature studies and were in accordance with the results of uniaxial testing performed on human soft tissues [7, 12, 15–23].

A small load of <0.1N was applied before the testing of each sample to ensure that there should not be slack before the test started. To ensure the accuracy and repeatability of the results, each sample was tested thrice on each strain rate. The test specimen size was 30 mm $\times$ 10 mm $\times$ 3 mm after clamping in the UTM grips, which was in line with literature studies [10]. For each experimental test, force–displacement data was collected and, after conversion, represented in terms of engineering stress–engineering strain plot. Figure 2 represents the experimental setup for the uniaxial testing of the test specimen.

Several steps were considered for processing raw force–displacement data to conclude the final results. Any part of the force–displacement curves with negative values was trimmed off and not considered for the analysis. The initial values of negative loads from positive displacements were excluded, and the plots were relocated to start from the origin.

2.4 Material Modeling of Stomach Tissue Surrogates

The human soft tissues and polymeric material (e.g., silicone-based polymer) exhibit non-linear mechanical behavior [24, 25]. The literature has reported that the stiffness of the human soft tissues changes with the variation in the strain rate [2, 10, 15,

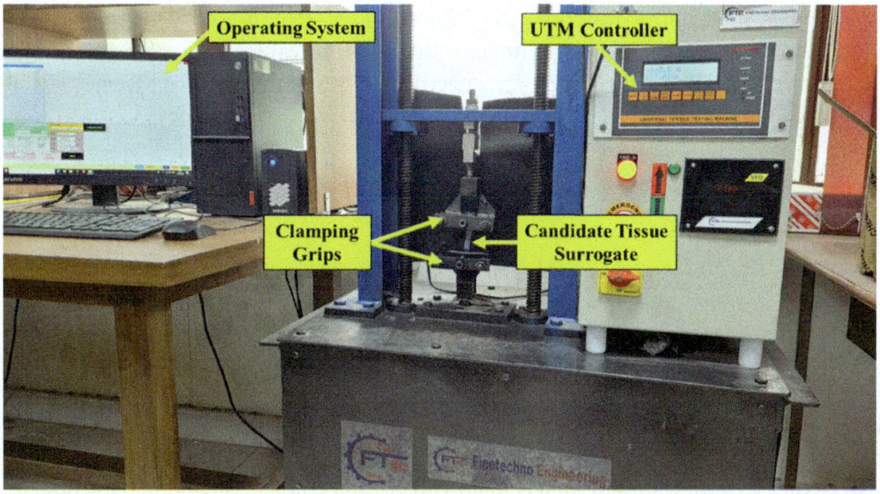

Fig. 2 Uniaxial tensile testing of candidate stomach tissue surrogates

26, 27]. Two hyperelastic curve fit models, namely Mooney–Rivlin model and the Yeoh model, were used to characterize the non-linear mechanical behavior of the developed stomach tissue surrogates. These hyperelastic formulation models depend on the strain energy function (Ψ) and vary depending on the material type [15, 28, 29]. Each hyperelastic curve fitting model also depends on Cauchy-Green tensor invariants *(I_1, I_2, and I_3)* or principal stretches *(λ_1, λ_2, and λ_3)* [28], as shown in Eqs. 1–4. As mentioned by Martins et al. [28], the principal Cauchy stress of soft materials can be expressed in the function of strain energy and the stretch values (Eq. 5):

$$\psi = \psi(I_1, I_2, I_3) \tag{1}$$

$$I_1 = \sum_{i=1}^{3} \lambda_i^2 \tag{2}$$

$$I_2 = \sum_{i,j=1}^{3} \lambda_i^2 \lambda_j^2 \tag{3}$$

$$I_3 = \prod_{i=1}^{3} \lambda_i^2 \tag{4}$$

$$\sigma_1 = \lambda_1 \frac{\partial \psi}{\partial \lambda_1} - \lambda_3 \frac{\partial \psi}{\partial \lambda_3}, \quad \sigma_2 = \sigma_3 = 0 \tag{5}$$

Using strain energy functions (Ψ) for Mooney–Rivlin and Yeoh hyperelastic models to solve Eq. (5), the experimental stress–strain values for each tested coupon were curve-fitted using Eqs. 6 and 7, respectively. The coefficients for Mooney–Rivlin (c_1 and c_2) and Yeoh (c_1, c_2, *and* c_3) models were evaluated in Sect. 3.

$$\sigma_{\text{Mooney-Rivilin}} = 2\left(\lambda^2 - \frac{1}{\lambda}\right)\left(c_1 + c_2\frac{1}{\lambda}\right) \tag{6}$$

$$\sigma_{\text{Yeoh}} = 2\left(\lambda^2 - \frac{1}{\lambda}\right)\left(c_1 + 2c_2(I_1 - 3) + 3c_3(I_1 - 3)^2\right) \tag{7}$$

Average R^2 correlation values of the tested coupons were considered for both hyperelastic models to determine the accuracy of the chosen hyperelastic models in predicting the material behavior of the stomach tissue surrogates. R^2 values vary from 0 to 1, where 0 represents the worst fit, and 1 shows the best fit. The human soft tissues exhibit regional material anisotropy and can be considered isotropic at a localized tissue level [2, 15, 27, 30]. Moreover, the literature studies reported tissue behavior to be isotropic or transversely isotropic when tested uniaxially under tensile loading conditions [31–33]. Therefore, our assumption that stomach tissue surrogates exhibit isotropic material behavior is reasonable.

3 Results and Discussion

3.1 Mechanical Testing of Candidate Stomach Tissue Surrogate Samples

The four-part polymeric material was arbitrarily mixed with varying parts A and B ratios in the specified proportion (as discussed in Sect. 2) to fabricate the test coupons of different shore hardness. Figure 3 shows the average stress–strain plot of all the experimental trials conducted at varying strain rates. The stress versus strain curves was plotted using the average stress values of each test coupon (i.e., 5A to 30A), with error bars representing the variation of stresses across all strain rates. The stress versus strain plots of the candidate samples for the stomach tissue surrogates was compared with the stress versus strain of the human stomach tissue from the literature [7].

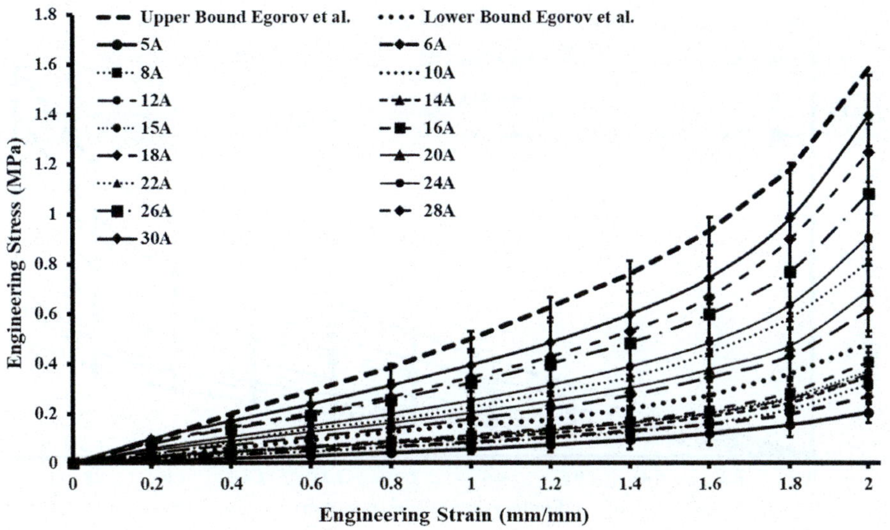

Fig. 3 Comparison of candidate stomach tissue surrogates with upper and lower bounds from the literature [7]

3.2 Repeatability Test of the Controlled Specimens

There were 15 test coupon candidates for stomach tissue surrogates, out of which 7 test coupons showed significant results. The results of these candidate test coupons were within the literature bounds for the human stomach tissue [7] and mimicked the mechanical properties of the stomach tissue (Fig. 4). The stress values (average ± S.D.) across all the six strain rates were found to be within the literature bounds for test coupons with shore hardness 18A, 20A, 22A, 24A, 26A, and 30A. The test coupons of shore hardness 18A and 30A were close to the lower and upper literature bounds, respectively, and identified as the control stomach tissue surrogates (Fig. 5) for the repeatability test.

The first control specimen was of shore hardness 30A (50%A-50%B) of the polymeric material. This control specimen closely mimics the upper bound of the literature stomach tissue [7] at strain rates of 0.8, 1, and 2.5 mm/s. For control specimen 1, five dedicated samples of shore hardness 30A were fabricated, and three sets of tests were conducted to validate the repeatability of the results (see Fig. 6). Similarly, for the second control specimen (i.e., 18A), 40%A-40%B of 15A shore hardness was mixed with 10%A-10%B of 30A to fabricate the five samples for the repeatability test. Three sets of tests were carried out with the five sample coupons, and the results of the repeatability test are shown in Fig. 7.

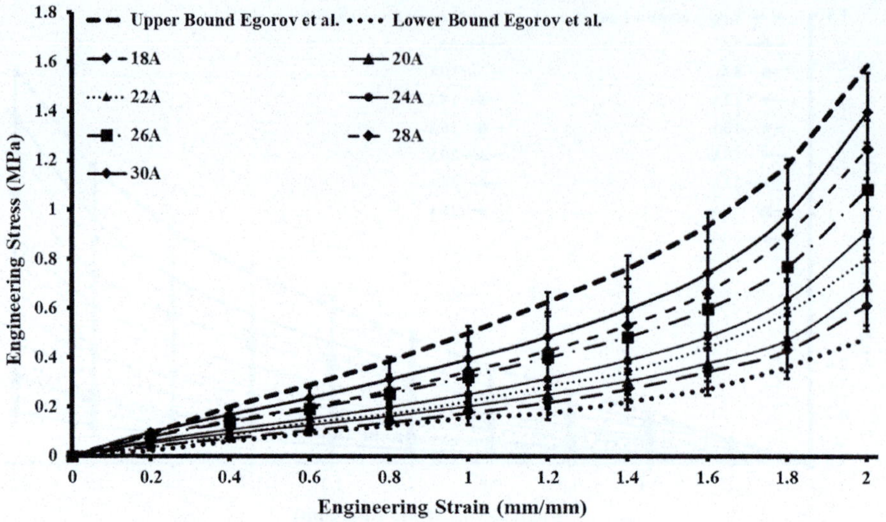

Fig. 4 Compositions of test coupons mimicking the mechanical behavior of stomach tissue and within the range of upper-lower bounds [7]

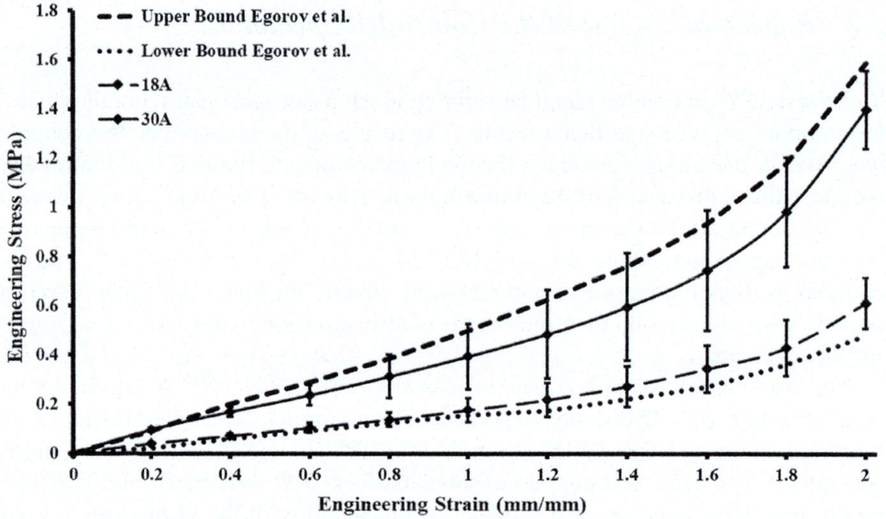

Fig. 5 Control stomach tissue surrogates mimicking the mechanical behavior of stomach tissue [7]

3.3 Hyperelastic Non-linear Curve Fits

Soft materials like polymers and human soft tissues have non-linear mechanical behavior that can be estimated using hyperelastic models [34–37]. In this work,

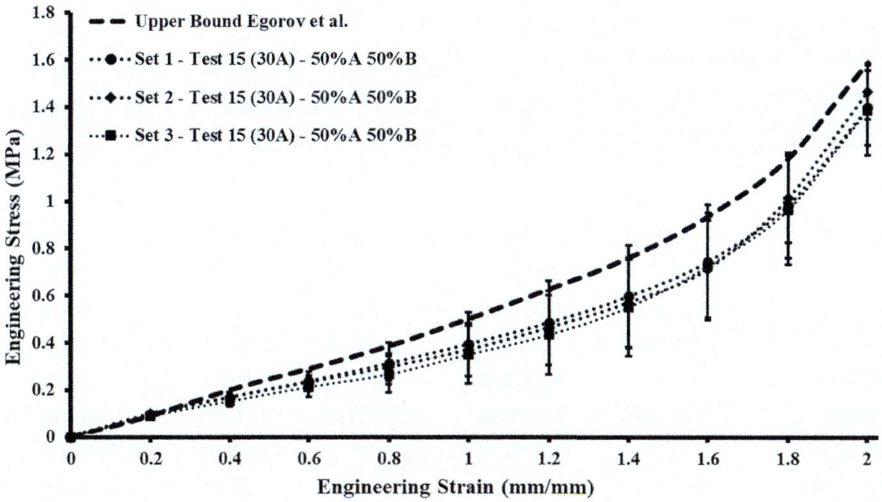

Fig. 6 Control specimen 1 (mean ± S.D.) mimicking the mechanical behavior of stomach tissue [7], tested for repeatability

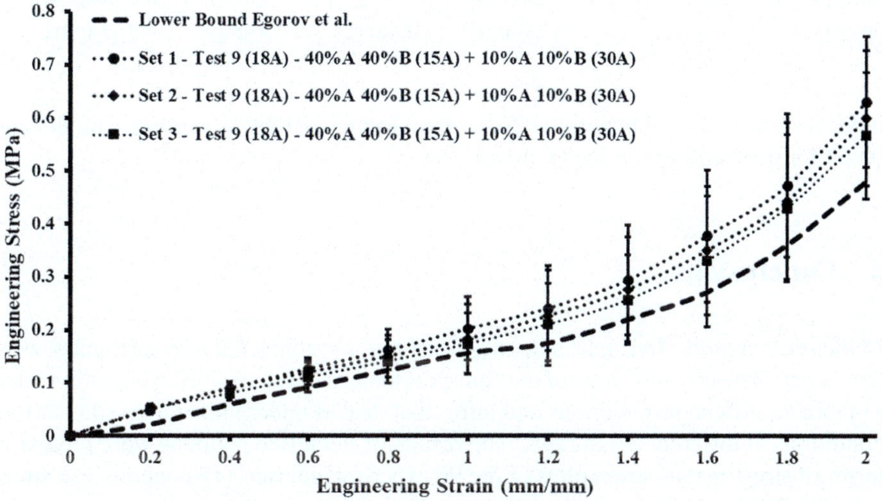

Fig. 7 Control specimen 2 (mean ± S.D.) mimicking the mechanical behavior of stomach tissue [7], tested for repeatability

Mooney–Rivlin and Yeoh hyperelastic models were used to evaluate the non-linear responses of the developed stomach tissue surrogates. As discussed in Sect. 2.4, Eqs. 6 and 7 were used to curve fit the experimental stress values of the uniaxially tested coupons. The estimated values of c_1 and c_2 for the Mooney–Rivlin model, and c_1, c_2, and c_3 for the Yeoh model are summarized in Table 2. Both the hyperelastic

Table 2 Hyperelastic coefficients for Mooney–Rivlin and Yeoh models

Hyperelastic curve fitting results					
Test coupon	Mooney–Rivlin		Yeoh model		
	c_1	c_2	c_1	c_2	c_3
Test 1	8.766E-3	4.873E-4	6.070E-3	4E-06	1.197E-05
Test 2	1.240E-2	6.513E-4	1.019E-2	6E-06	2.918E-05
Test 3	1.346E-2	6.928E-4	9.580E-3	1.1E-05	6.043E-05
Test 4	1.765E-2	8.467E-4	1.434E-2	1.4E-05	3.784E-05
Test 5	1.260E-2	6.713E-4	1.353E-2	1.7E-05	4.719E-05
Test 6	1.529E-2	7.231E-4	1.278E-2	1.9E-05	5.157E-05
Test 7	1.629E-2	8.147E-4	1.385E-2	2.3E-05	5.654E-05
Test 8	1.728E-2	8.699E-4	1.396E-2	2.6E-05	6.578E-05
Test 9	2.822E-2	9.698E-4	2.267E-2	2.8E-05	8.544E-05
Test 10	3.073E-2	9.947E-4	2.559E-2	3.1E-05	9.074E-05
Test 11	3.900E-2	1.072E-3	2.872E-2	3.7E-05	1.250E-04
Test 12	4.025E-2	1.169E-3	3.184E-2	4E-05	1.403E-04
Test 13	4.715E-2	1.215E-3	4.126E-2	4.3E-05	1.417E-04
Test 14	5.570E-2	1.297E-3	4.358E-2	5.9E-05	1.988E-04
Test 15	6.065E-2	1.489E-3	4.961E-2	6.5E-05	2.115E-04

models showed $R^2 > 0.9$, with 0.925 for the Mooney–Rivlin hyperelastic model and 0.964 for the Yeoh hyperelastic model.

4 Conclusions

In the present work, four-part silicone-based human stomach tissue surrogates were developed. The compositions of the four-part material system were arbitrarily varied to fabricate fifteen test coupons and, after that, tested under uniaxial tensile loading conditions at different strain rates. The results of the tested coupons were plotted in terms of stress versus strain plots of the 90 experimental runs (15 coupons $\times$ 6 strain rates) and compared with the literature on cadaveric stomach tissue to determine the material compositions mimicking the mechanical properties of the stomach tissue. Mooney–Rivlin and Yeoh hyperelastic curve fit models were used to characterize the non-linear behavior of the developed stomach tissue surrogates. The R^2 value of 0.925 for Mooney–Rivlin and 0.964 for the Yeoh models shows the proficiency of the selected hyperelastic curve fit models. Across all the tested compositions, two test coupons closely mimicked the mechanical properties of the stomach tissue and were identified as controlled samples. The controlled samples were tested thrice, and the output results confirm the repeatability of the tissue surrogates.

There are a few limitations of the four-part silicone-based material system, which should be acknowledged. First, the developed material system exhibited isotropic properties and was unable to incorporate the anisotropic behavior. It was reported in the literature that human soft tissues show isotropic and transversely isotropic behavior under uniaxial tensile loading conditions. Therefore, the considered limitation was not of much concern in the present work. Second, it was found that there was an insignificant effect of variation in strain rates up to the strain of 1. In the case of cadaveric tissues, strain rates were reported to affect the stress versus strain plots even at small strains (e.g., <1). However, a significant variation was found with an increase in the strain, i.e., high strain >1.

The developed stomach tissue surrogates possess no ethical and biosafety concerns like the testing/handling of cadaveric tissue samples in clinical testing. Such tissue surrogates would be indispensable for surgical training, educational purposes, and studying wide medical models and biomechanics of stomach tissue.

Acknowledgements Gurpreet Singh is grateful to the Ministry of Education, Government of India, for awarding the Prime Minister's Research Fellowship (Ref: IITD/Admission/Ph.D./PMRF/2020-21/4062) for pursuing his doctoral research program at IIT-Delhi, India.

Conflict of Interest Statement The authors declare no conflict of interest with respect to the research, authorship, and/or publication of this article.

References

1. Brandstaeter S, Fuchs SL, Aydin RC, Cyron CJ (2019) Mechanics of the stomach: A review of an emerging field of biomechanics. GAMM-Mitteilungen 42. https://doi.org/10.1002/gamm.201900001
2. Singh G, Chanda A (2021) Mechanical properties of whole-body soft human tissues: a review. Biomed Mater 16:062004. https://doi.org/10.1088/1748-605X/AC2B7A
3. Aydin RC, Brandstaeter S, Braeu FA, Steigenberger M, Marcus RP, Nikolaou K et al (2017) Experimental characterization of the biaxial mechanical properties of porcine gastric tissue. J Mech Behav Biomed Mater 74:499–506. https://doi.org/10.1016/j.jmbbm.2017.07.028
4. Zhao J, Liao D, Gregersen H (2005) Tension and stress in the rat and rabbit stomach are location- and direction-dependent. Neurogastroenterol Motil 17:388–398. https://doi.org/10.1111/j.1365-2982.2004.00635.x
5. Zhao J, Liao D, Chen P, Kunwald P, Gregersen H (2008) Stomach stress and strain depend on location, direction and the layered structure. J Biomech 41:3441–3447. https://doi.org/10.1016/j.jbiomech.2008.09.008
6. Rosen J, Brown JD, De S, Sinanan M, Hannaford B (2008) Biomechanical properties of abdominal organs in vivo and postmortem under compression loads. J Biomech Eng 130. https://doi.org/10.1115/1.2898712
7. Egorov VI, Schastlivtsev IV, Prut EV, Baranov AO, Turusov RA (2002) Mechanical properties of the human gastrointestinal tract. J Biomech 35:1417–1425. https://doi.org/10.1016/S0021-9290(02)00084-2
8. Lim YJ, Deo D, Singh TP, Jones DB, De S (2009) In situ measurement and modeling of biomechanical response of human cadaveric soft tissues for physics-based surgical simulation. Surg Endosc 23:1298–1307. https://doi.org/10.1007/s00464-008-0154-z

9. Standard Test Method for Rubber Property—Durometer Hardness n.d. https://www.astm.org/d2240-15e01.html. Accessed 21 Jan 2022
10. Chanda A (2018) Biomechanical modeling of human skin tissue surrogates. Biomimetics 3:18. https://doi.org/10.3390/BIOMIMETICS3030018
11. Chanda A, Unnikrishnan V, Lackey K, Robbins J (2019) Biofidelic conductive soft tissue surrogates. 69:127–35. https://doi.org/10.1080/00914037.2018.1552856.
12. Makode S, Singh G, Chanda A (2021) Development of novel anisotropic skin simulants. Phys Scr 96:125019. https://doi.org/10.1088/1402-4896/AC2EFD
13. Singh G, Gupta V, Chanda A (2022) Artificial skin with varying biomechanical properties. Mater Today Proc. https://doi.org/10.1016/J.MATPR.2022.03.433
14. Chanda A, Callaway C (2018) Tissue anisotropy modeling using soft composite materials. Appl Bionics Biomech. https://doi.org/10.1155/2018/4838157
15. Chanda A, Callaway C, Clifton C, Unnikrishnan V (2018) Biofidelic human brain tissue surrogates. Mech Adv Mater Struct 25:1335–1341. https://doi.org/10.1080/15376494.2016.1143749
16. Bisplinghoff JA, Kemper AR, Duma SM (2012) Dynamic material properties of the pregnant human uterus. J Biomech 45:1724–1727. https://doi.org/10.1016/J.JBIOMECH.2012.04.001
17. Martins PALS, Filho ALS, Fonseca AMRMI, Santos A, Santos L, Mascarenhas T, et al (2011) Uniaxial mechanical behavior of the human female bladder. Int Urogynecol J 22:991–995. https://doi.org/10.1007/s00192-011-1409-0
18. Karimi A, Shojaei A (2018) An Experimental Study to Measure the Mechanical Properties of the Human Liver. Dig Dis 36:150–155. https://doi.org/10.1159/000481344
19. Bourgouin S, Bège T, Masson C, Arnoux PJ, Mancini J, Garcia S et al (2012) Biomechanical characterisation of fresh and cadaverous human small intestine: applications for abdominal trauma. Med Biol Eng Comput 50:1279–1288. https://doi.org/10.1007/S11517-012-0964-Y/FIGURES/6
20. Karimi A, Shojaei A, Tehrani P (2017) Measurement of the mechanical properties of the human gallbladder. J Med Eng Technol 41:541–545. https://doi.org/10.1080/03091902.2017.1366561
21. Jin X, Zhu F, Mao H, Shen M, Yang KH (2013) A comprehensive experimental study on material properties of human brain tissue. J Biomech 46:2795–2801. https://doi.org/10.1016/J.JBIOMECH.2013.09.001
22. Kemper AR, Santago AC, Stitzel JD, Sparks JL, Duma SM (2012) Biomechanical response of human spleen in tensile loading. J Biomech 45:348–355. https://doi.org/10.1016/j.jbiomech.2011.10.022
23. Polio SR, Kundu AN, Dougan CE, Birch NP, Ezra Aurian-Blajeni D, Schiffman JD et al (2018) Cross-platform mechanical characterization of lung tissue. PLoS ONE 13:e0204765. https://doi.org/10.1371/JOURNAL.PONE.0204765
24. Chanda A, Unnikrishnan V, Roy S, Richter HE (2015) Computational modeling of the female pelvic support structures and organs to understand the mechanism of pelvic organ prolapse: a review. Appl Mech Rev 67. https://doi.org/10.1115/1.4030967/370016
25. Shergold OA, Fleck NA, Radford D (2006) The uniaxial stress versus strain response of pig skin and silicone rubber at low and high strain rates. Int J Impact Eng 32:1384–1402. https://doi.org/10.1016/J.IJIMPENG.2004.11.010
26. Chanda A, Graeter R (2018) Human skin-like composite materials for blast induced injury mitigation. J Compos Sci 2:44. https://doi.org/10.3390/jcs2030044
27. Chanda A, Flynn Z, Unnikrishnan V (2018) Biomechanical characterization of normal and prolapsed vaginal tissue surrogates. J Mech Med Biol 18. https://doi.org/10.1142/S0219519417501007
28. Martins PALS, Natal Jorge RM, Ferreira AJM (2006) A comparative study of several material models for prediction of hyperelastic properties: application to silicone-rubber and soft tissues. Strain 42:135–147. https://doi.org/10.1111/j.1475-1305.2006.00257.x
29. Holzapfel GA (2000) Nonlinear solid mechanics : a continuum approach for engineering 455
30. Prange MT, Margulies SS (2002) Regional, directional, and age-dependent properties of the brain undergoing large deformation. J Biomech Eng 124:244–252. https://doi.org/10.1115/1.1449907

31. Velardi F, Fraternali F, Angelillo M (2005) Anisotropic constitutive equations and experimental tensile behavior of brain tissue. Biomech Model Mechanobiol 51:53–61. https://doi.org/10.1007/S10237-005-0007-9
32. Meaney DF (2003) Relationship between structural modeling and hyperelastic material behavior: application to CNS white matter. Biomech Model Mechanobiol 1:279–93. https://doi.org/10.1007/S10237-002-0020-1
33. Christ AF, Franze K, Gautier H, Moshayedi P, Fawcett J, Franklin RJM et al (2010) Mechanical difference between white and gray matter in the rat cerebellum measured by scanning force microscopy. J Biomech 43:2986–2992. https://doi.org/10.1016/J.JBIOMECH.2010.07.002
34. Singh G, Chanda A (2023) Development and biomechanical testing of artificial surrogates for vaginal tissue. Adv Mater Process Technol. https://doi.org/10.1080/2374068X.2023.2198837
35. Singh G, Chanda A (2023) Biofidelic gallbladder tissue surrogates. Adv Mater Process Technol. https://doi.org/10.1080/2374068X.2023.2198835
36. Singh G, Chanda A (2023) Development and mechanical characterization of artificial surrogates for brain tissues. Biomed Eng Adv 5:100084. https://doi.org/10.1016/J.BEA.2023.100084
37. Gupta V, Singh G, Chanda A (2022) Development and testing of skin grafts models with varying slit orientations. Mater Today Proc 62:3462–3467. https://doi.org/10.1016/J.MATPR.2022.04.282

Chapter 8
Hip Joint Prosthesis Using SiC CMC and Ti-6Al-4V Materials

Anshul Tripathi, Sahil Thakur, and Tushar Aggarwal

1 Introduction

Hip joint prostheses, also known as hip arthroplasty, are a commonly performed surgical procedure to replace damaged or diseased hip joints to restore their normal function [1–3]. The selection of material for hip joint prostheses is crucial as it significantly impacts the biocompatibility, wear resistance, and mechanical properties of the prosthesis [4, 5]. The success of hip joint prosthesis depends on the selection of an appropriate material that closely mimics the mechanical properties and biological characteristics of natural bone [6–8]. Conventional metallic materials such as cobalt-chromium and titanium alloys have been traditionally used in hip joint prostheses [9–11]. However, these materials are associated with some drawbacks such as corrosion, wear, and fatigue failure [12]. Additionally, these traditional materials have the potential to induce adverse reactions in the human body, such as metal sensitivity, allergic reactions, and systemic toxicity due to the release of metallic ions caused by corrosion [13]. Ceramic matrix composites (CMCs), reinforced by ceramic fibers and embedded in a ceramic matrix, have emerged as a potential alternative to traditional materials due to their superior biocompatibility, wear resistance, and mechanical properties [14]. Silicon carbide (SiC) fiber-reinforced silicon carbide matrix (SiC) CMC has been considered a material for hip joint prostheses due to its exceptional mechanical properties and wear resistance [15]. SiC CMC has been reported to have a high strength-to-weight ratio, high modulus of elasticity, high thermal conductivity, and excellent wear resistance [16]. However, the use of SiC CMC in hip joint prostheses remains limited, and further research is necessary to evaluate its suitability as a material for hip joint prostheses [17]. Ti-6Al-4V, an alloy composed primarily of titanium (90–95%) with small amounts of aluminum and vanadium, has been

A. Tripathi (✉) · S. Thakur · T. Aggarwal
Department of Mechanical Engineering, Manav Rachna University, Faridabad, Haryana 121004, India
e-mail: anshul.tripathi@icloud.com

A. Chanda et al. (eds.), *Materials for Biomedical Simulation*, Materials Horizons: From Nature to Nanomaterials, https://doi.org/10.1007/978-981-99-5064-5_8

used in a wide range of applications, such as aircraft and automotive components, medical implants, and sporting goods [18]. Aluminum provides strength and hardness to the alloy, while vanadium improves its ductility and toughness. The small amount of iron added to Ti-6Al-4V increases its strength [19]. Ti-6Al-4V is widely used due to its combination of strength and lightness, but it also has some limitations, including difficulty in machining and susceptibility to cracking and corrosion [13, 14]. In conclusion, the selection of material for hip joint prostheses is a critical factor that affects the success of the procedure. While SiC CMC and Ti-6Al-4V both have their advantages and limitations, further research is necessary to evaluate the suitability of SiC CMC as a material for hip joint prostheses.

The aim of this study is to investigate the mechanical performance of hip joint prostheses made of SiC CMC and Ti-6Al-4V using Finite Element Analysis (FEA) as shown in Fig. 1. The procedure involves creating a 3D CAD model of the prosthesis using SolidWorks software and importing it into ANSYS for the FEA analysis. The model has undergone various loading conditions, including static loads that imitate body weight, walking, and climbing stairs. The FEA results provide an understanding of the stress and strain distribution throughout the prosthesis under different loading conditions, as well as shed light on the mechanical behavior of SiC CMC and Ti-6Al-4V. This study also compares the results with literature values of bone to determine the suitability of SiC CMC as a material for hip joint prostheses. Based on the findings, recommendations for future research in this field have been presented.

Fig. 1 Hip joint prosthesis

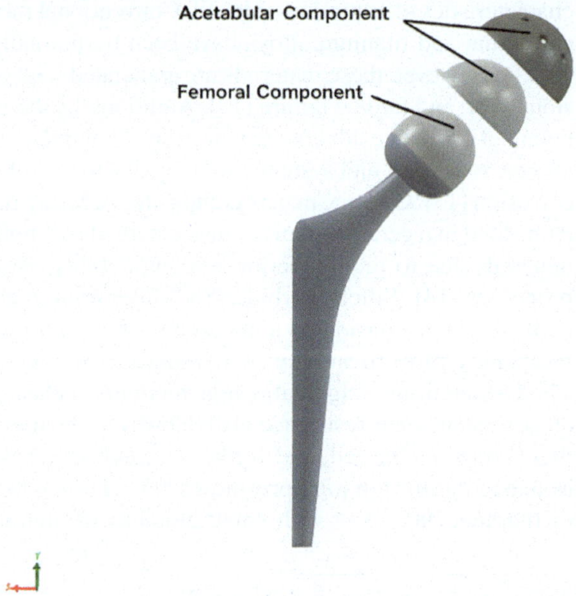

2 Geometric and Material Specification

The selection of an appropriate material for hip joint prostheses is critical in ensuring optimal biocompatibility, wear resistance, and mechanical properties. After conducting a comprehensive literature review and evaluation of various materials, the study has determined that the Silicon Carbide fiber-reinforced Silicon Carbide matrix (SiC) ceramic matrix composite (CMC) is the most suitable material for the hip joint prosthesis as presented in Table 1. These advanced materials are recognized for their exceptional mechanical characteristics, including high strength-to-weight ratio, high modulus of elasticity, high thermal conductivity, and exceptional wear resistance. It should be noted that Ti-6Al-4V, an alloy primarily composed of titanium (90–95%) with small amounts of aluminum and vanadium, is a commonly utilized material in hip joint prostheses. Despite its widespread use due to its balance of strength and lightness, Ti-6Al-4V has some limitations such as difficulties in machining and a tendency towards cracking and corrosion. The geometric specifications of the hip joint prosthesis, including the precise dimensions and shapes of the femoral and acetabular components, play a crucial role in ensuring proper fit and functionality. These components must comply with ISO 7206 standards and possess precise tolerances, surface finishes, and geometric accuracy.

The specific geometric specifications will be determined based on the size and shape of the patient's femur and hip socket for optimal performance and to minimize the risk of dislocation as illustrated in Fig. 2. In conclusion, the study has deemed SiC CMC to be the most suitable material for the hip joint prosthesis due to its exceptional mechanical properties. However, the use of Ti-6Al-4V in hip joint prostheses is also widely accepted and its limitations should be considered during the material selection process.

Table 1 Material properties of SiC-CMC, Ti-6Al-4V, and bone

Materials	Properties			
	Density (gm/cm^3)	Young's modulus (GPa)	Yield strength (MPa)	Poisson's ratio
SiC CMC	3.2	400	800	0.17
Ti-6Al-4V	4.43	120	900	0.34
Bone	1	7–30	50–150	0.3

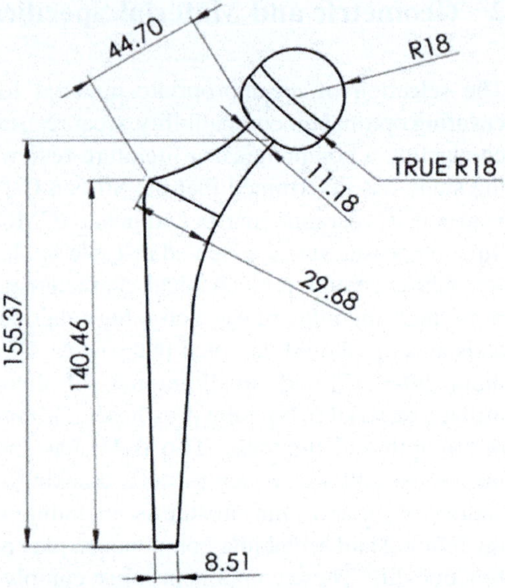

Fig. 2 Geometric specifications (mm) of hip joint prosthesis

3 Methodology

3.1 Mesh Topology

The generation of a mesh is a vital aspect in the implementation of finite element analysis (FEA) for the evaluation of a hip joint prosthesis. The mesh, consisting of small units known as finite elements, serves as a computational representation of the prosthesis geometry. The finite elements approximate the prosthesis's response to various loading conditions through this representation. The process of mesh generation begins with the development of a computer-aided design (CAD) model of the prosthesis, which outlines its geometry. Subsequently, meshing software is utilized to convert the CAD model into finite elements via the creation of a three-dimensional mesh, as depicted in Fig. 3.

The size, shape, and distribution of these elements must be selected carefully to ensure that they are fine enough to capture important details of the prosthesis, but coarse enough to be computationally manageable. The mesh generation process is crucial in ensuring the validity and reliability of the FEA results.

3.2 Boundary Conditions

In this study, the finite element analysis (FEA) of a hip joint prosthesis was performed to assess its mechanical behavior. To accurately capture the real-world behavior of

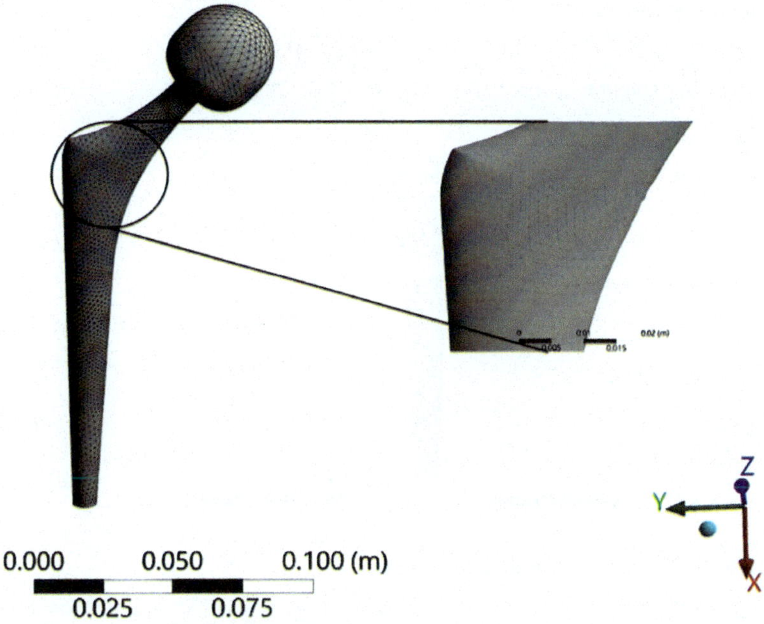

Fig. 3 Mesh generation of hip joint prosthesis

the prosthesis, the application of kinematic constraints and loading conditions was of utmost importance. The femoral component of the prosthesis was fixed to constrain its movement in one or more degrees of freedom, while a load was imposed on the acetabular component to simulate the loading scenarios encountered during activities such as walking or stair-climbing as shown in Fig. 4. It is crucial to note that the proper implementation of kinematic constraints and loading conditions is essential for obtaining reliable and accurate results from the FEA, and for ensuring that the prosthesis will perform optimally in vivo. The precise application of these constraints and conditions is therefore a critical aspect of this study, and crucial for achieving its objectives.

4 Results and Discussion

In order to evaluate the performance of the materials under different loading conditions, stress and displacement analyzes were conducted for body weight (250N), walking (350N), and stair-climbing (450N) using finite element analysis (FEA). The results for all three materials, SiC-CMC, Ti-6Al-4V, and Bone, were analyzed and compared.

The stress values obtained from the stress analysis are an indication of how well the materials can withstand external loading conditions. Higher stress values indicate

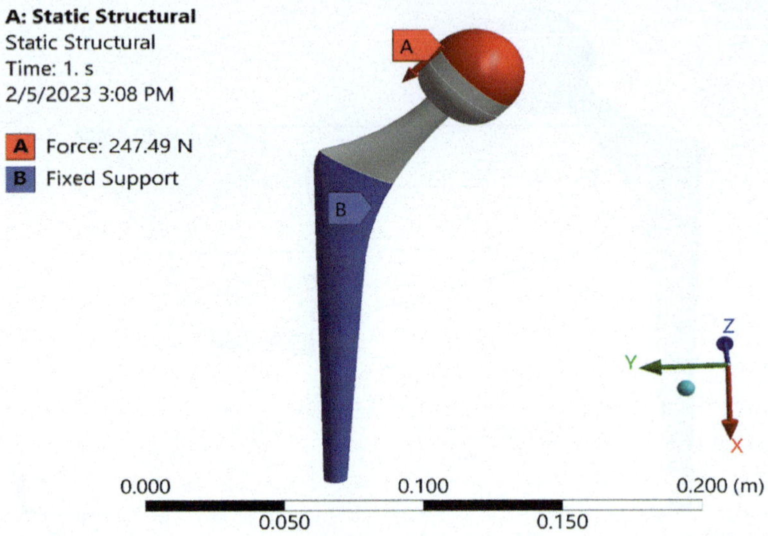

Fig. 4 Boundary conditions for hip-joint prosthesis

that the material is experiencing greater internal stresses, which may increase the risk of material failure or deformation. Therefore, lower stress values are generally preferred for materials used in load-bearing applications such as hip joint prostheses.

In the case of the stress analysis results for the three materials (Bone, SiC-CMC, and Ti-6Al-4V) under a 250N load condition, the values obtained for SiC-CMC and Bone are within a similar range, indicating that SiC-CMC has comparable stress-bearing properties to bone. On the other hand, the stress value for Ti-6Al-4V is slightly lower than the other two materials, suggesting that it may be less suitable for use in load-bearing applications than SiC-CMC and Bone.

The displacement analysis results in Fig. 6 for Bone, SiC-CMC, and Ti-6Al-4V under a 250N load provide insights into the materials' ability to deform and recover their shape under load. The values indicate that SiC-CMC has the lowest displacement value of 4.80E-04 mm, which suggests that it has a higher stiffness and is less likely to deform under load compared to Bone and Ti-6Al-4V. In contrast, the displacement values for Bone and Ti-6Al-4V are relatively higher at 1.10E-02 mm and 3.60E-03 mm, respectively, indicating that they are more prone to deformation under load and may require additional support or reinforcement in a prosthetic device.

The displacement analysis is an important complement to the stress analysis, as it provides information on the material's ability to recover its shape and resist permanent deformation, which is important for maintaining the longevity and functionality of the prosthetic device. Therefore, the combination of stress and displacement analysis provides a more complete understanding of the material behavior under different loading conditions and can inform the selection and design of materials for hip joint prostheses.

The stress values for all three materials are relatively close under the 350N load, with SiC-CMC exhibiting the highest stress value. This suggests that SiC-CMC may experience higher levels of stress than Bone and Ti-6Al-4V when subjected to walking-like loading conditions.

The displacement values for all three materials are significantly different under the 350N load. SiC-CMC exhibits the lowest displacement value, indicating that it is more rigid than Bone and Ti-6Al-4V under walking-like loading conditions. This information can be used to better understand the mechanical behavior of these materials and inform the design of hip joint prostheses.

Figure 9 shows the stress (MPa) distribution for Bone, SiC-CMC, and Ti-6Al-4V under a load of 450N. The values of stress for Bone, SiC-CMC, and Ti-6Al-4V are 14.46, 15.38, and 14.2, respectively. The results indicate that SiC-CMC exhibits higher stress values compared to the other materials under this loading condition. However, it should be noted that all materials show stress values within an acceptable range for use as hip joint prostheses. These findings demonstrate the importance of conducting stress analysis under different loading conditions to ensure the safety and efficacy of the prosthetic material.

Figure 10 shows the displacement (in mm) of Bone, SiC-CMC, and Ti-6Al-4V under a load of 450N. The displacement values for Bone, SiC-CMC, and Ti-6Al-4V were 2.04E-02, 8.84E-04, and 6.65E-03, respectively. The results suggest that SiC-CMC has the lowest displacement values, indicating that it is stiffer and can better resist deformation under load. This property is desirable in a hip joint prosthesis as it can help maintain the integrity of the implant-bone interface and minimize the risk of implant loosening or failure. The higher displacement values for Bone and Ti-6Al-4V may indicate a higher risk of implant loosening or failure under load.

As shown in Figs. 5, 6, 7, 8, 9 and 10, the stress and displacement values for each material varied significantly depending on the loading condition. For example, for the body weight condition, SiC-CMC exhibited the lowest maximum stress value of 27.9 MPa, while Ti-6Al-4V exhibited a maximum stress value of 66.5 MPa. However, for the stair-climbing condition, Ti-6Al-4V exhibited the lowest maximum stress value of 81.4 MPa, while SiC-CMC exhibited a maximum stress value of 117.8 MPa. Similarly, the displacement values also varied significantly between the materials for different loading conditions.

Overall, the stress and displacement analyzes provided valuable insights into the performance of the materials under different loading conditions as shown in Table 2. The results suggest that the selection of material for hip joint prostheses should depend on the specific loading conditions it will be subjected to. For example, as illustrated in Figs. 11, 12, if the patient is likely to engage in activities that require higher loading conditions, Ti-6Al-4V may be a more suitable material due to its better performance under stair-climbing conditions. However, for less demanding activities, SiC-CMC may be a better option due to its lower stress values for body weight and walking conditions.

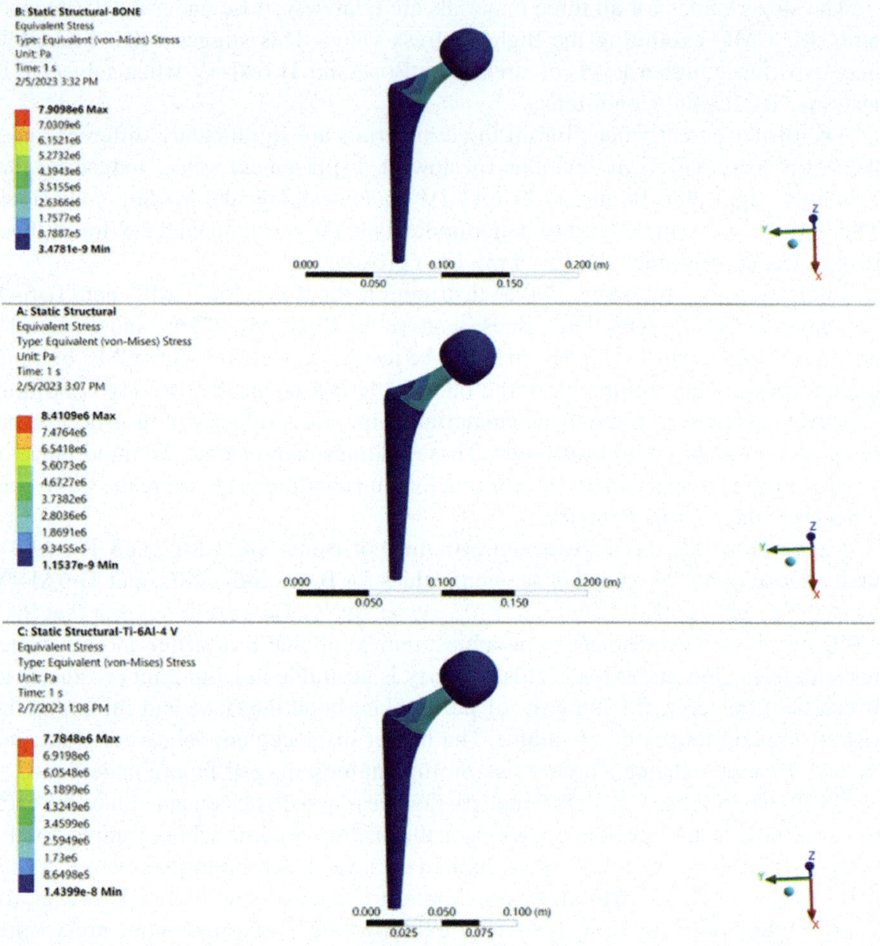

Fig. 5 Stress under 250N load for bone, SiC-CMC, and Ti-6Al-4V

In summary, the stress and displacement analyzes showed that the performance of the materials varied significantly depending on the loading conditions. The results provide valuable information for the selection of materials for hip joint prostheses and can help in optimizing their design for better performance and longevity.

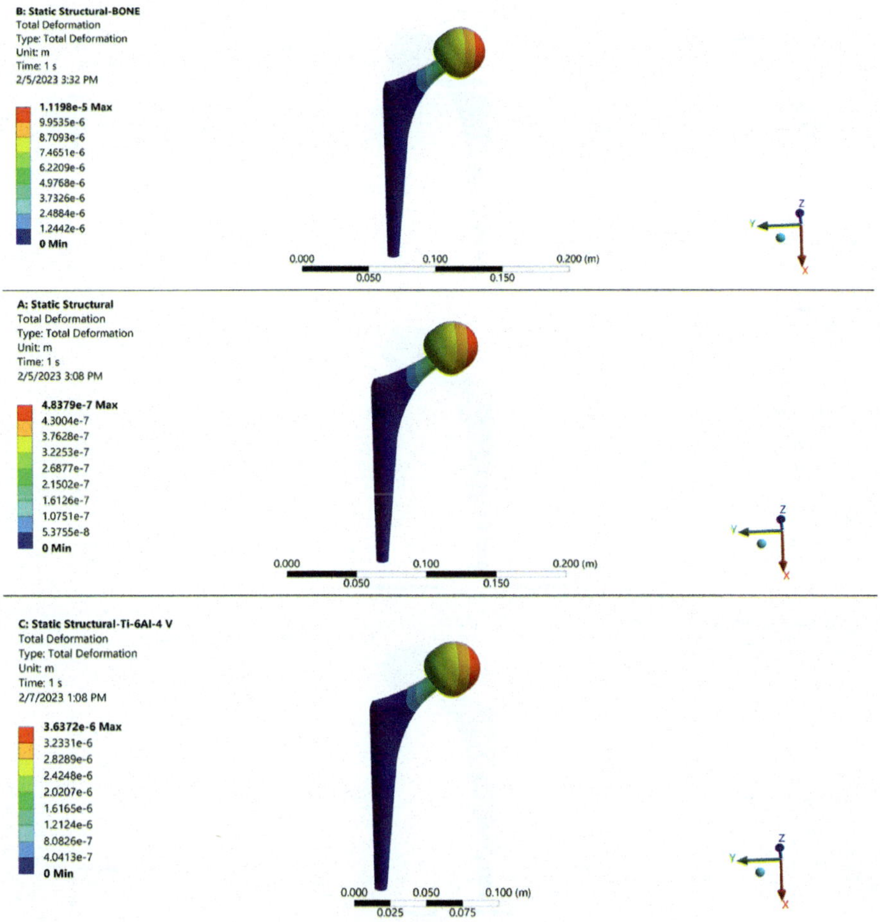

Fig. 6 Displacement under 250N load for bone, SiC-CMC, and Ti-6Al-4V

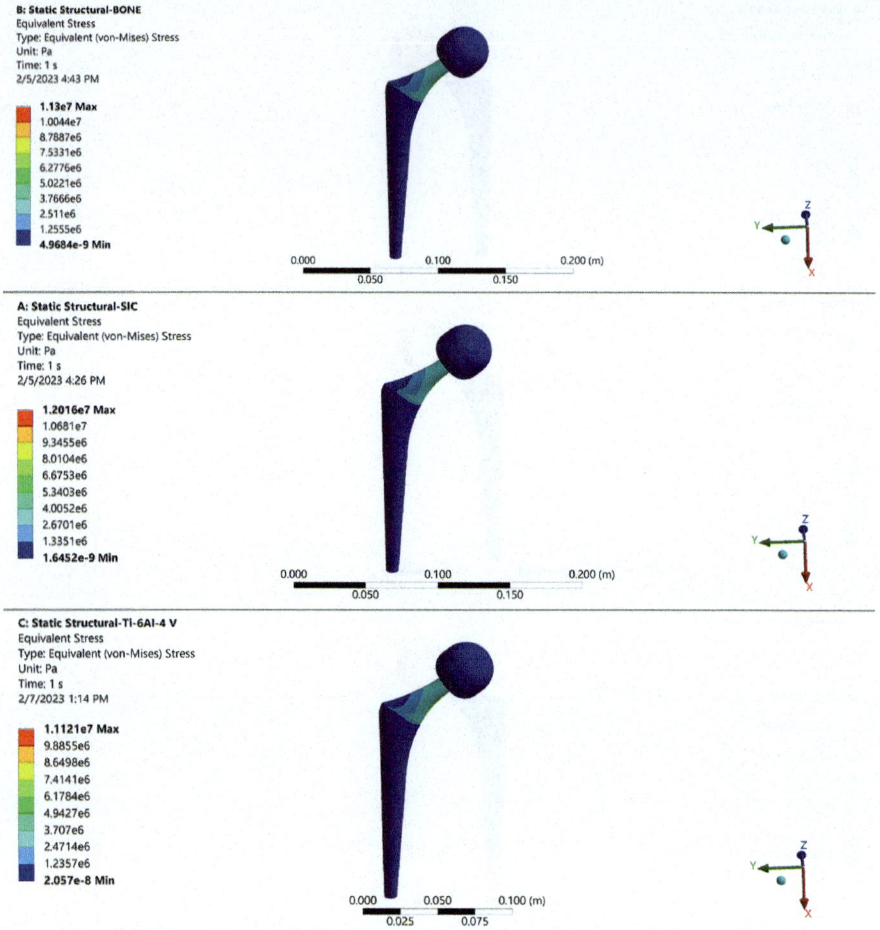

Fig. 7 Stress under 350N load for bone, SiC-CMC, and Ti-6Al-4V

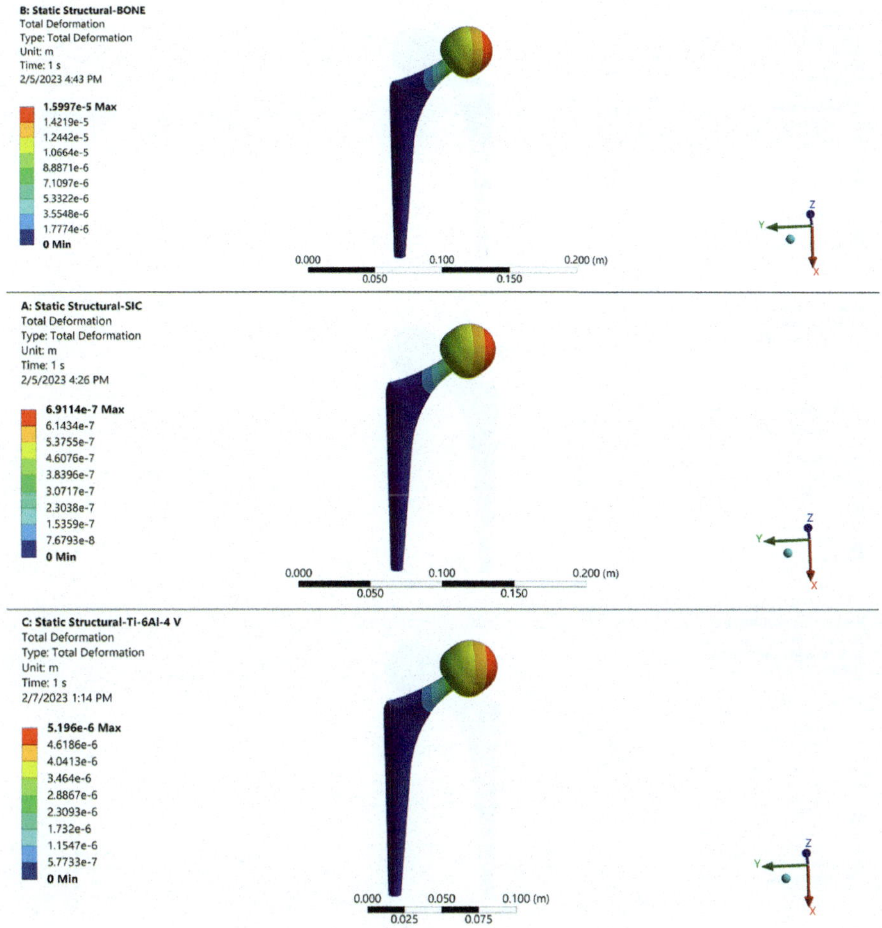

Fig. 8 Displacement under 350N load for bone, SiC-CMC, and Ti-6Al-4V

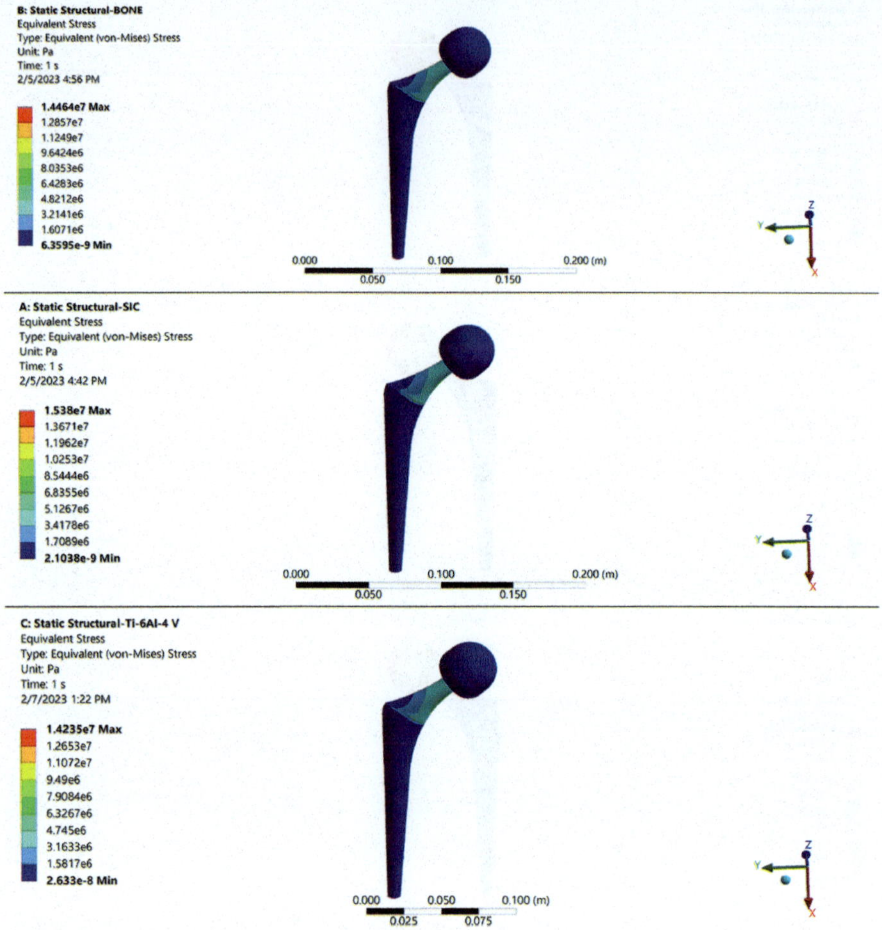

Fig. 9 Stress under 450N load for bone, SiC-CMC, and Ti-6Al-4V

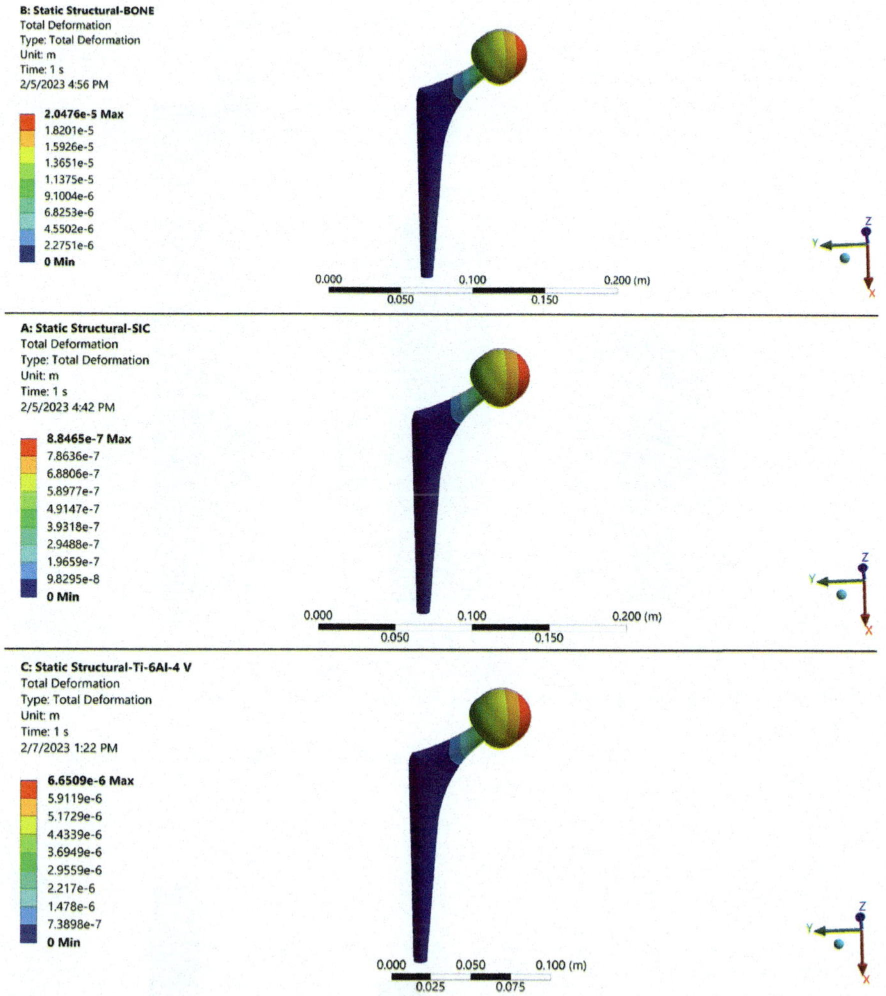

Fig. 10 Displacement under 450N load for bone, SiC-CMC, and Ti-6Al-4V

Table 2 Comparison of generated stresses and displacements

Force (N)	Description	SiC CMC	Bone	Ti-6Al-4 V
250	Stress (MPa)	8.4	7.9	7.78
	Displacement (mm)	4.80E-04	1.10E-02	3.60E-03
350	Stress (MPa)	12.01	11.3	11.12
	Displacement (mm)	6.90E-04	1.59E-02	5.19E-03
450	Stress (MPa)	15.38	14.46	14.2
	Displacement (mm)	8.84E-04	2.04E-02	6.65E-03

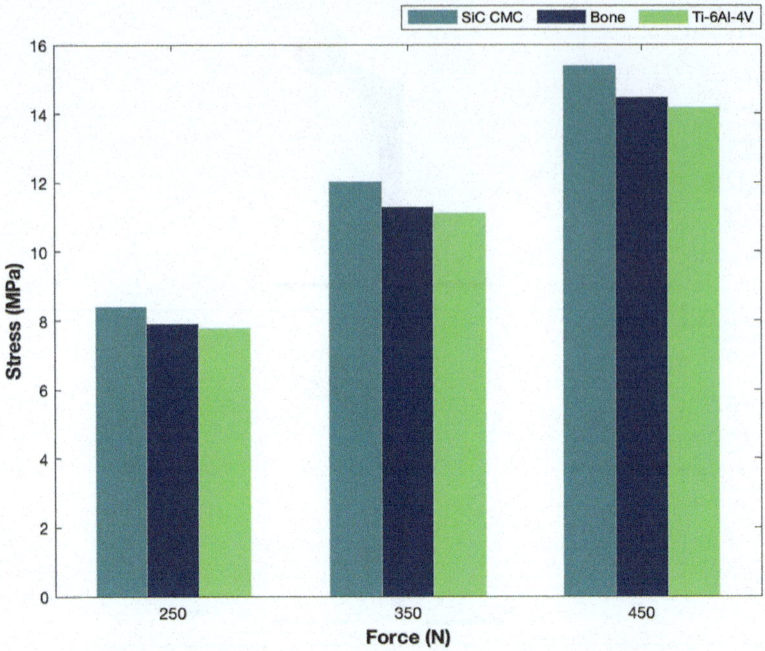

Fig. 11 Plot between force and stress for SiC-CMC, Bone, and Ti-6Al-4V

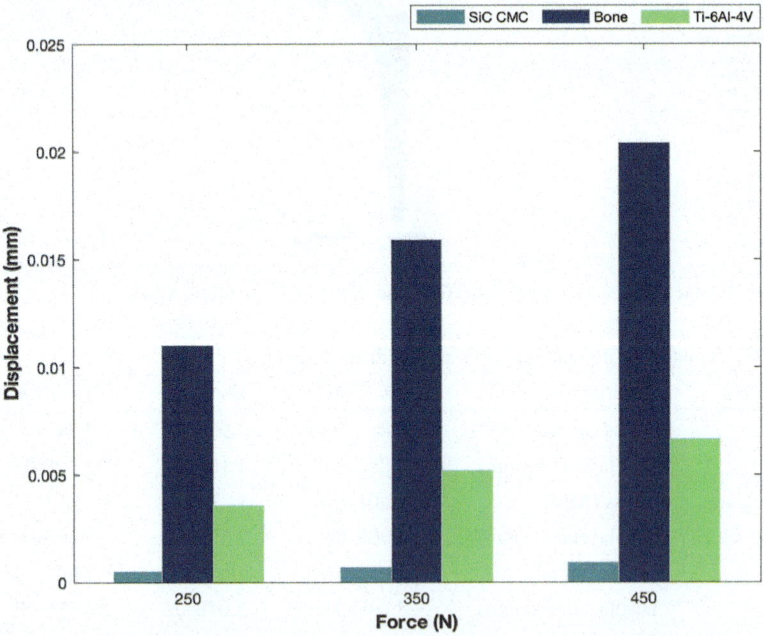

Fig. 12 Plot between force and displacement for SiC-CMC, Bone, and Ti-6Al-4V

5 Conclusion

In conclusion, this research aimed to evaluate the performance of various materials for use in hip joint prostheses. The findings of this study indicate that Silicon Carbide fiber-reinforced Silicon Carbide matrix (SiC) ceramic matrix composite (CMC) and Ti-6Al-4V both exhibit desirable properties for use in hip joint prostheses. However, after a thorough literature review and evaluation, SiC CMC was chosen as the most beneficial material due to its exceptional mechanical properties, such as high strength-to-weight ratio, high modulus of elasticity, high thermal conductivity, and superior wear resistance. The results of the stress and displacement analysis show that SiC CMC is comparable to the mechanical properties of natural bone, making it a suitable choice for use in hip joint prostheses. These findings have important implications for the design and development of future hip joint prostheses, as they indicate that SiC CMC has the potential to provide superior performance and durability compared to other materials.

References

1. Lewinnek GE, Lewis JL, Tarr R, Compere CL, Zimmerman JR (1978) Dislocations after total hip-replacement arthroplasties. Arthroplast J Bone Jt Surg-Am 60(4):427–429
2. Singh G, Sidhu SS, Bains PS, Bhui AS (2019) Improving microhardness and wear resistance of 316L by TiO2 powder mixed electro-discharge treatment. Mater Res Express 6(8):086501
3. Ablyaz TR, Shlykov ES, Muratov KR (2020) Surface characterization and tribological performance analysis of electric discharge machined duplex stainless steel. Micromachines 11(10):926
4. Bhui AS, Singh G, Sidhu SS, Bains PS (2018) Experimental investigation of optimal ED machining parameters for Ti-6Al-4V biomaterial. Facta Univ Ser Mech Eng 16(3):337–345
5. Nargol AV, Glyn-Jones S, Gardiner DJ (2015) Wear of metal-on-metal hip replacements. J Bone Joint Surg 97(7):569–576
6. Zhang WJ, Ries MD, Cui FJ (2010) The wear of orthopaedic bearing couples. J Mech Behav Biomed Mater 3(5):749–759
7. Smith LG, Meneghini RM (2011) Total Hip Arthroplasty. J Bone Joint Surg 93(1):24–32
8. Singh G, Lamichhane Y, Bhui AS, Sidhu SS, Bains PS, Mukhiya P (2019) Surface morphology and microhardness behavior of 316L in HAp-PMEDM. Facta Univ Ser Mech Eng 17(3):445–454
9. Yang KH, Kim SH, Kim JK, Park JH (2015) Wear of ultra-high molecular weight polyethylene and metal-on-metal in total hip arthroplasty. J Orthop Sci 20(1):87–93
10. Szomor Z, Adám G, Balogh L (2008) Complications and failures of hip joint replacement. Orv Hetil 149(45):2139–2146
11. Liu Y, Fan J, Li H, Wang Q (2016) Characteristics and applications of ceramic matrix composites in biomedical engineering. Mater Sci Eng, C 61:196–208
12. Papp JF, Hulka GW, Karch J (2007) Silicon carbide fiber-reinforced silicon carbide matrix composites for biomedical applications. J Mater Sci Mater Med 18(1):1–9
13. Bains PS, Bahraminasab M, Sidhu SS, Singh G (2019) On the machinability and properties of Ti–6Al–4V biomaterial with n-HAp powder–mixed ED machining. Proc Inst Mech Eng Part H J Eng Med 234(2):232–242

14. Singh G, Sidhu SS, Bains PS, Singh M, Bhui AS (2020) On surface modification of Ti alloy by electro discharge coating using hydroxyapatite powder mixed dielectric with graphite tool. J Bio Tribo-Corrosion 6(3):91
15. Akhras O, Bouzid J, Goidanich M (2018) Influence of the fiber orientation on the mechanical properties of SiC fiber reinforced SiC matrix composites. Int J Refract Metal Hard Mater 74:56–62
16. Hsu TW, Chen YS, Lo SL (2006) Wear behaviour of SiC fiber-reinforced SiC matrix composites. Wear 261(5–6):744–752
17. Matthews FL, Jakus P (2015) Ti-6Al-4V titanium alloy: Properties, processing, and applications. Mater Sci Eng R Rep 96:1–45
18. Chandra R (2017) Ti-6Al-4V titanium alloy: characteristics, properties, and applications. In: Titanium alloys in medical applications. Woodhead Publishing, pp 1–22
19. O'Dowd NP, Cleary PJ (2015) Ti-6Al-4V: the workhorse of titanium alloys. J Mater Eng Perform 24(3):896–903

Chapter 9
Applications of Nano Materials in Dental Sciences and Scope in Future Practice

Mohammad Afazal and Saba Afreen

1 Introduction to Nanotechnology

Richard Feynman first proposed the idea of nanotechnology in his famous speech "There's Plenty of Room at the Bottom" [1]. It attempts to use molecular engineering to create nano machineries that can produce nano materials. Under "Definition of Nanotechnology," an early NNI document (National Science and Technology Council [NSTC], 2000) claimed that the ability to operate at the molecular level, atom by atom, to develop enormous structures with fundamentally new molecular organization is the essence of nanotechnology. The behavior of structural features in the range of 10^{-9} to 10^{-7} (1–100 nm) exhibits significant alterations when compared to the behavior of isolated molecules of approximately 1 nm (10^{-9}) or of bulk materials. We will be able to organize atoms in any way we like, thanks to nanotechnology, which will allow us to effectively and completely control the structure of matter [2, 3].

Nanomedicine is a brand-new area that has emerged as a result of developments in the applications of nanotechnology in medicine [4]. According to Robert A. Freitas Jr., who first proposed this idea in 1993, this notion entails using nanostructures and nanodevices to observe, manage, and treat the biological systems of the human body at the molecular level [5]. The pharmaceutical business, where many medications are hydrophobic in nature and only sparsely or weakly soluble in aqueous solutions, is one of the most straightforward applications of nanotechnology. Due to the enormous increase in surface area, shrinking a pharmaceutical down to the nanoparticle size range can enable more of the drug to go into the solution [6]. Dental treatments using designed nano materials can improve therapeutic and preventative outcomes.

M. Afazal (✉)
University Polytechnic, Jamia Millia Islamia Delhi, New Delhi 110025, India
e-mail: mafazal1@jmi.ac.in

S. Afreen
Indraprastha Dental College and Hospital, Sahibabad, Ghaziabad 201010, Uttar Pradesh, India

A. Chanda et al. (eds.), *Materials for Biomedical Simulation*, Materials Horizons: From Nature to Nanomaterials, https://doi.org/10.1007/978-981-99-5064-5_9

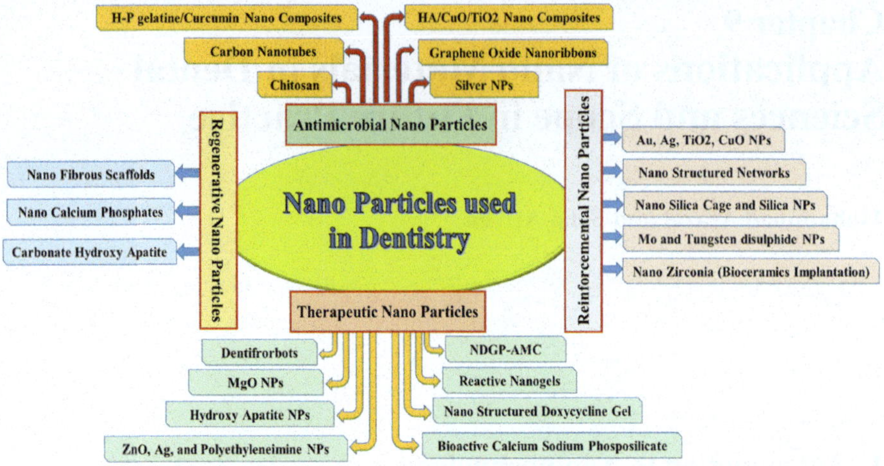

Fig. 1 Classes of nano materials used in dentistry

2 Nano Dentistry

The most recent advances in nanotechnology research and development can be used to improve and upgrade conventional dental materials and equipment. In order to create regenerative materials, nano fillers, nano composites, nano impression materials, target-oriented antimicrobial mouthwashes, implants, and drug-enclosing nanoparticles, various synthesis approaches, including bottom up, top down, functional, and biomimetic approaches, can be used in nano dentistry. A thorough grasp of the tooth's structure and environment, as well as expertise in the material, synthesis process, and synthesis technique, are necessary for the development of nano dental products. Since ancient times, dental biomaterials made of metals, ceramics, resins, and polymers have been created and further modified [7]. Polymers can be made from both synthetic and natural (plant- and animal-based) materials [8]. To improve the mechanical and optical properties of polymers, some uncommon materials, such as carbon nanotubes (CNT), graphene oxide, and nanodiamonds, can be used in place of traditional glass or carbon fibers [9] (Figs. 1 and 2).

3 Applications in Diagnosis

For the prevention or treatment of oral disorders caused by biofilms, a thorough understanding of bacterial adhesion—the primary factor in bacterial colonization and pathogenesis—as well as bacterial nano mechanics is necessary [10]. The ability of bacteria to cling to individuals of the same or other species as well as to various substrates, such as teeth and implants, has been well documented [11, 12]. AFM offers

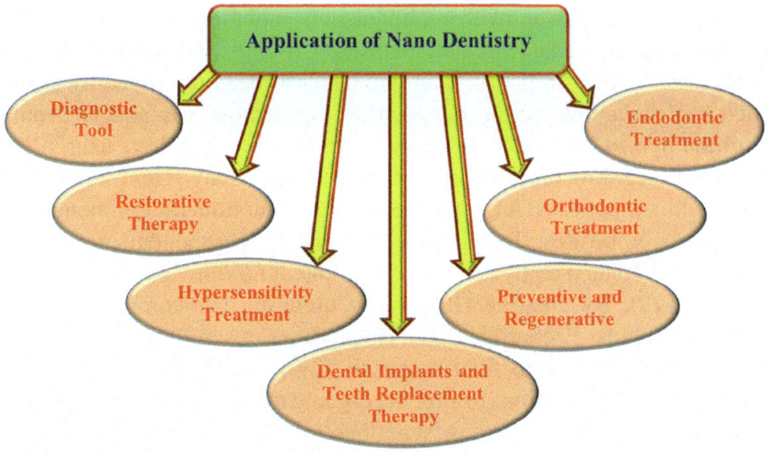

Fig. 2 Applications of nanoparticles in dentistry

a breakthrough in the characterization of bacteria as well as the measurement of their adhesion to various substrates [13, 14], because of its capacity to directly interact with imaging live cells without affecting their morphology and properties [15]. A real-time, highly sensitive scanning of a living bacterial cell was made possible with nanomechanical biosensors and an AFM cantilever [16]. Additionally, details on the characteristics of the membrane molecules [17] and the elasticity of a cell [18] are made available.

4 Application of Nano Dentistry in Conservative and Restorative Treatment of Teeth

The primary treatment needed in the oral cavity is caries management. To preserve the tooth's structural and cosmetic integrity, an appropriate restorative material has to be used to fill the tooth. After a tooth is lost, it is restored using a bridge, an implant, and a crown. Nanoparticles can be added to filling materials and implants to improve them. The material must be close to the tooth structure and adhere to it in order for the restoration of a carious tooth to be successful. Micro gaps, leaks, and ultimately filling failure are caused by compromised adhesion or intersurface integrity [19, 20]. For enhanced defense, mineral deposition, and sealing of exposed collagen fibers in prepared teeth, the use of nano fillers or nano gels has been tried [21]. Reactive nanogels [22], zirconia (20–50 nm) [23], HA (20–70 nm) [24], colloidal silica (5–40 nm) or barium aluminosilicate nanofillers (400 nm) [25], and bioactive calcium/sodium phosphosilicate [21] are some examples of nano materials. Self-healing adhesives that can patch up microscopic or nanoscale cracks without compromising the strength of resin-dentin connections have also been developed [26]. They contain nano capsules that are packed with healing agents, and the presence of a break in the resin matrix typically causes the nano capsules to burst and release their contents. The durability of the resin-dentin interface has also been improved by using nano-controlled molecular interaction between the resin monomer and HA that is still present in the hybrid layer [27]. In comparison to fine size fillers, nano fillers offer superior adherence, biocompatibility, and durability. Many scientific organizations are using hydroxyapatite (HA), the primary inorganic mineral component of teeth, to create nano composites. To serve as an inorganic component of nano composites, Chung et al. created a physiological component mixture using HA and chitosan nanoparticles [28]. By creating HP-gelatin/curcumin nanocomposites, antimicrobial properties against Escherichia coli, Staphylococcus aureus, and Streptococcus mutans were introduced into filler materials [29]. Various natural polymers have been employed in investigations to alter the organic nanocomposites' organic component characteristics [30]. The strongest antibacterial HA/CuO/TiO_2 nanocomposites were created by Imani et al. The strong Quality by Design principle assisted in the manifestation of development [31]. A difficult challenge to incorporate Ag NPs, graphene oxide (GO), multi-walled carbon nanotubes (MWCNTs), and graphene oxide nanoribbons (GONRs) in HA nanocomposites was recently undertaken by Balu et al. The team investigated how the amount of carbon affected the final hardness of the nanocomposites. Additionally, a bio strain was used to assess the Ag NPs activity (E. coli and S. aureus bacteria). Lidocaine was made available as a model drug release by the created nanocomposite [32].

As a supplementary material in the creation of nanocomposites, other synthetic polymers have been investigated. Poly Methyl Methacrylate Nanocomposites have improved mechanical properties as compared to HA nanocomposites [33]. Mallakpour et al. developed multi-component nanocomposites (Polyvinylpyrrolidone/L-leucine Amino Acid, Functionalized Mg-Substituted Fluorapatite Nanocomposites) using ultrasonic waves as an energy source [34]. Microwaves [35–37], photons [38] and their combinations [39] are other energy sources that have been used by various research groups to create nano composites.

Open tubules are brushed with highly concentrated GNPs to prevent pain and suffering in dentinal hypersensitivity, and laser irradiation is used to encourage the aggregation of nanoparticles to cover the exposed tubules [40]. Additionally, dental nanorobots provide a rapid and effective treatment for dentin hypersensitivity by accurately and selectively occluding the tubules with biological materials in a matter of minutes [4].

5 Applications in Endodontics Treatment

The innermost, most important, and vascular component of a tooth is the dental pulp, which is made up of nerve fibers and blood vessels. The area of dentistry known as endodontics deals with periodontal tissues and dental pulp. When an infection penetrates the dentin and enamel to reach the pulp, pulp treatment is necessary. Cleaning and preparing root canals that contain pulp tissues for an appropriate filling and sealing material is the focus of endodontics. The following qualities should be present in an ideal root-end filling material: a hermetic seal, non-resorb ability, non-toxicity, non-carcinogenicity, biocompatibility, and dimensional stability [41–43]. None of the materials in existence satisfies every requirement. To sanitize root canals, root canal sealers contain a number of nanoparticles, including zinc oxide and chitosan alone or in combination. The flow characteristics of the sealers were unaffected, but they improved the antibacterial action, as evidenced by a notable decrease in Enterococcus faecalis adhering to treated dentin [44]. It was possible to retain the inhibitory impact of a chitosan-modified root canal sealer on biofilm growth at the sealer-dentin interface by first treating the surface of root canal dentin with phosphorylated chitosan [45]. Metal oxides, including magnesium oxide nanoparticles, show promised antibacterial activity in both in vitro and ex vivo tests and may be employed as a possible root canal irrigant. Magnesium oxide nanoparticles (5 mg/L) show a statistically significant long-term benefit in the removal of E as compared to the usual NaOCl solution (5.25%). Faecalis adhered to the dentin of the root canal [46].

Nano materials that aim to regenerate pulp tissue could improve endodontic treatment by preserving the health of the pulp and, by extension, the structural integrity of the tooth. Dentin Odontoblast (located at the edge of the pulp) stem cells have also been investigated for this purpose; however, the clinical setup showed a futile response [47, 48]. Hanafy et al. investigated two widely used dental biomaterials, namely mineral trioxide aggregate (MTA) and nano-HA, as odontogenic differentiation promotors. The results showed significantly higher and upregulated expression of the odontotomy differentiation-specific genes, namely OPN, RUNX2, OCN, and Collagen1, in the treatment group compared to the control group [49]. Another strategy to keep the pulp healthy and keep the infection outside the vascular area is pulp capping. Li et al. suggested a combination of micro-nano bioactive glasses with biocompatible, osteogenesis-sensitizing characteristics. Dental pulp capping using Ca–Zn–Si-based micro nanospheres (Zn doped). The results were positive, showing increased antibacterial effects and increased macrophage stimulation to decrease proinflammatory indicators, followed by dentin remineralization via sensitization of dental pulp cells [50]. Drug-encapsulated liposomes were suggested by Sinjari et al. as a very sophisticated nanotechnological method for restoring the homeostasis of dental pulp stem cells. In terms of 2-hydroxyethyl methacrylate, the treatment was able to help restore cell proliferation and reduce inflammation markers [51]. Kim et al. have suggested an RGD peptide conjugated dendrimer-based medication delivery system for dental pulp differentiation following severe dental injury. Higher mineralization and odontogenic potential were very positive findings [52]. With a very small number of tests, Elgendy and Fayyad investigated natural scaffolds, such as propolis and chitosan, for tooth restoration and discussed their potential for endodontic treatment because of their high biocompatibility and capacity for tissue restoration [53]. In a similar vein, Tondnevis et al. suggested creating a dental tissue scaffold utilizing polymers and the freeze-drying method that contains nano-HA or Nano-Fluro HA/Chitosan scaffold. Results showed that chitosan was helpful in significantly increasing cell proliferation [54]. These experiments demonstrated the chitosan NPs' potential for use in dental endodontics. The role of eggshell-derived porous nano-HA and CMC (Carboxy Methylcellulose) composite was recently reported in a laboratory study by Baskar et al., which demonstrated their impact on dental bioactivity and cell proliferation using the significantly increased levels of VEGF and dentine sialo phosphoprotein [55]. The development and testing of amoxicillin-loaded nanodiamond Gutta-percha composite (NDGP-AMC) for use in root canal procedures produced encouraging results [56]. Gelatin [57–59], collagen [60], silk [59], and other natural fibers and polymers have also been investigated in nano dentistry.

6 Applications in Orthodontic Treatment

To realign teeth, force must be applied in the desired direction and interfacial stability must exist between the tooth surface, bracket, arch wire, and ligatures, among others [61]. Better holding and stability are prevented by the frictional forces between the arch wire and the brackets, which can also lengthen the course of treatment and have an impact on the final result. These components are covered with Fullerene, such as Molybdenum and tungsten disulfide NPs, to lessen this friction [61]. Also, nanotechnology can help in maintaining better oral hygiene, better anchorage, and lesser enamel demineralization during orthodontic treatment by several coatings like elastomeric ligatured supported NPs (Benzocaine and Ag) [62], nanocomposites (Ag NPs with ZnO, chlorhexidine) of adhesive types/bands [63], gold NPs in orthodontic adhesives [64], silver NPs [65], TiO_2 NPs [66], and copper oxide NPs [67].

7 Applications in Prosthodontic Treatment

Patients are increasingly more likely to seek out rehabilitation therapy as a result of greater awareness of and interest in quality-of-life enhancement. Nanotechnology is now a component of materials for crown, bridge, and implants used in restorative dentistry as a result of progress in research aimed at improving tooth replacement. The qualities of currently employed materials, including ceramics, impression materials, denture bases, and different forms of prosthodontic cement, have also been greatly improved by nanotechnology. Polymethylmethacrylate (PMMA) polymers are mostly used in removable prosthodontic appliances such as complete dentures, partial dentures, and detachable maxillofacial prostheses. Chlorinated polyethylene, poly(methyl methacrylate), poly(urethane), poly(vinyl chloride), and poly(dimethylsiloxane) (PDMS—silicone elastomers) have all been employed in the production of maxillofacial prosthesis [68]. Although PMMA has strong biocompatibility, aesthetics, processability, and reparability, it has the drawbacks of being weak, having a low fracture resistance, behaving radiopacitively, and having microbial adherence

[69–71]. Well-dispersed nano-ZrO_2 particles can increase the strength, modulus, and ductility while TiO_2 nanoparticles strengthened the mechanical behavior of PMMA. Particles of Ag TiO_2 and Fe_2O_3 considerably lessen C's adhesion. albicans made of PMMA that have no impact on growth or metabolism [72–74]. It can aid in treating pathological conditions such as denture stomatitis brought on by Candida albicans adhering to the denture base materials [75]. Due to their sophisticated mechanical and physical qualities, silicone elastomers have been employed in the production of dental prosthesis for a long time [76]. They are also non-toxic, chemically resistant, and biocompatible. PDMS [77, 78] and silicone rubber are the most preferred types [79–81]. In recent studies, various filler nanoparticles (NPs) have been added to silicone rubbers to enhance their physical and mechanical properties through [82]. The silicone matrix is most frequently reinforced using silica NPs as fillers [82, 83]. Silicone elastomers' mechanical and physical properties would vary depending on the concentration of silica NPs present [82, 84]. Additionally, silicone elastomers are supported by polyhedral oligomeric silsesquioxanes (POS), a nano (1.5 nm) silica cage, and metal nanoparticles (NPs) to enhance their tensile strength and other physical characteristics. Fixed prosthesis can use nano materials, nanocomposites, and nano coatings. Cytotoxicity brought on by the leaching of organic monomers can be resolved by nanocomposites built on nanofiller technology. To increase flexural strength and hardness, 3-methacryloxypropyl-tri-methoxy-silane was applied to silica particles to create zirconia-silica nanoparticles. Nanotechnology offers new opportunities and promises for enhancing the adherence and endurance of implants. In addition to calcium phosphate, silica-based NPs, polyvinyl alcohol, and carbon nanotubes were employed to create nanocomposites and occasionally scaffolds for enhancing mechanical strength and tissue regeneration [85]. The most recent developments in research on the uses of nanometals, nanoceramics, nanoresins, and other nano materials in prosthodontics have been reviewed. This research clearly demonstrates that many properties of materials used in prosthodontics, such as modulus elasticity, surface hardness, polymerization shrinkage, and filler loading, can be significantly improved after their scales were reduced from micron-size into nano size by nanotechnology.

8 Application in Dental Implants

Nanostructure-modified titanium implants encourage osteogenic differentiation and could have a better bio-integration into the alveolar bone [86]. The flat surface of titanium implants can be anodized to create nanotubular structures with a diameter of less than 100 nm [87]. The physicochemical characteristics of surfaces [88], as well as the spacing and diameter of nanotubes, can be controlled by altering variables including voltage, current density, and the chemistry of the electrolyte. Long nano tube arrays (10 m) and pillar-like nanostructures with adjustable sizes are also deposited by anodization on titanium surfaces [89]. On Titanium, Ti6Al4V, Cr–Co–Mo alloys, and Tantalum, nano pit networks (pit diameter 20–100 nm) can be successfully created by combining strong acids or bases with oxidants. According to a study, treatment with HCl produces superior outcomes than those with H_2SO_4 and $Na_2S_2O_8$ [90]. Acid etching can be used in conjunction with other procedures, such as grit blasting, to remove contaminants on implant surfaces by blasting residues. The titanium dental implants' osseointegration could be hampered by the grit blasting residue. But compared to an acid-etched surface, the nanostructured Ti surface created by physical vapor deposition had a surface area increase of up to 40% and a stronger osseointegration. Additionally, to improve bone regeneration, HA nanocrystals [91] and calcium-phosphorus NPs [92] can be used to treat or modify the implant surface. A wide range of distinctive nanostructures, including octahedral bipyramids, nano flowers, nano needles, nano rods, and meso-porous nano scaffolds, have been produced on titanium using a combination of hydrothermal treatments (tuning concentration, temperature, reaction medium composition, and time duration) and sodium hydroxide [93]. Early bone healing and improved mechanical interlocking with bone result from the deposition of discrete 20–40 nm nanoparticles on a dual acid-etched implant surface. Niobium oxide and diamond-like carbon nano topographies have been produced on titanium and other substrates with the aid of a mixture of chemical vapor deposition and the sol–gel method, improving the bioactivity of implantable metals. Other current techniques for creating a nanostructured dental implant surface include laser technology and coatings made of ultraviolet (UV) photo functionalized (picometer to nanometer) TiO_2 [94].

9 Applications Preventive Therapy

Anticaries DNA vaccine's immunogenicity has been improved by the use of customized delivery vehicles, such as anionic liposomes in chitosan/DNA nanoparticle complexes. In order to permit the release of the vaccine in a pH-dependent way, the surface charge of the delivery vehicle may also be pH-dependent. All in vitro research up to this point has suggested that nanotechnology may be able to stop the progression of early caries lesions in their surface but not deeper layers. A local application of a nanostructured doxycycline gel has been employed in an experimental periodontal disease model to stop bone loss. With their continuous and quick movement (1–10 m/s) across the supra and subgingival surfaces, nanorobots (dentifrobots) mouthwash or toothpaste left on the occlusal surfaces of teeth continually remove the organic residues and prevent the calculus accumulation. When eaten, these nanorobots can be safely deactivated [95]. A number of oral health care products, including liquids and pastes containing nano-appetite for managing biofilm at the tooth surface and goods including nano materials for remineralizing early submicrometric sized enamel defects, have been developed using biomimetic techniques. On the tooth surface, bacteria form biofilms that lead to dental cavities. Nanocomposite surface coatings can minimize bacterial adhesion, prevent pathogenic effects, and make the tooth surface easier to clean. The apatite nanoparticle-containing toothpaste can be employed as biofilm management nano materials and as a method for remineralizing enamel lesions that are smaller than sub-micrometers [96–98]. However, these oral nanoparticle preventive medications are still in the research phase, and further in-depth research is required before they can be used in clinical settings. Several nanoparticles, like zinc oxide, silver, and polyethyleneimine, can be incorporated into dental composites or dental adhesives to give antibacterial nano treatment. This inhibits the growth of germs through a variety of ways. The bacterial cell membrane is disrupted, sugar metabolism and active transport are both inhibited, reactive oxygen species are produced, magnesium ions necessary for the enzymatic activity of oral biofilms are displaced, electron transport across the bacterial membrane is disturbed, and DNA replication is prevented [99, 100].

10 Applications in Regenerative Therapy

Different types of nano-calcium phosphates, including dicalcium phosphate anhydrous, tetra calcium phosphate, monocalcium phosphate monohydrate, and carbonate hydroxyapatite, have been employed as Ca- and PO-releasing fillers for remineralization in recurrent caries [101, 102]. Similar to other medical fields, tissue engineering in dentistry has been used to integrate scaffold matrices with the regeneration abilities of stem cells, which are mostly derived from dental tissues such as dental pulp, periodontal ligament (PDL), and alveolar bone. These scaffolding materials have been significantly improved thanks to nanotechnology in tissue engineering, creating special 3D matrix conditions for cells and tissues. A native tissue architecture can be developed using a bottom up method, giving an engineered construct the mechanical properties of enamel and dentin [103]. The dental tissues' nanoarchitecture is used to create electro spun nanofibers, self-assembling peptides, and phase-separation matrices. The development of nanofibrous scaffolds as matrices for the regeneration of dental tissues, such as the dentin-pulp complex, enamel, PDL, cementum, alveolar bone, and temporomandibular joint, has been widespread [104, 105].

11 Conclusion

Nano dentistry attracts patients to dentistry because it is cost-effective, saves time, and prevents psychological trauma. More patient-centered research will aid in the advancement of nanotheranostics that are both effective and cost-effective. Despite the numerous irrefutable gaps that limit its clinical exploration, revolutionary nanotechnology has enhanced conventional dentistry. Research in the field of nano dentistry still lags behind other areas of biological study. Prior to the application of nanotechnology on a large scale, fundamental molecular engineering methods, mass production techniques, and the simultaneous coordination of many nanorobots must be overcome. Nanotechnology advancements are shaping the future of healthcare administration. They have the potential to produce significant benefits, such as improved health, more efficient use of natural resources, and less environmental pollution. Nanotechnology will profoundly alter dentistry, healthcare, and human life. However, social issues of public acceptance, ethics, regulation, and human safety must be addressed prior to the incorporation of molecular nanotechnology into the modern medical and dental arsenal.

References

1. Drexler KE (2004) Nanotechnology: from Feynman to funding. Bull Sci Technol Soc 24(1):21–27. https://doi.org/10.1177/0270467604263113
2. Mansoori GA (2005) Advances in atomic and molecular nanotechnology. In: Principles of nanotechnology, pp. 1–30. https://doi.org/10.1142/9789812569318_0001
3. Rieth M (2003) Nano-engineering—studies and conclusions. In: Nano-engineering in science and technology, pp. 91–132. https://doi.org/10.1142/9789812560032_0005
4. Freitas RA (2000) Nanodentistry. J Am Dent Assoc 131(11):1559–1565. https://doi.org/10.14219/jada.archive.2000.0084
5. Robert A Nanomedicine. Basic capabilities, vol I, 1st edn. https://www.routledge.com/Nanomedicine-Volume-I-Basic-Capabilities/Freitas/p/book/9781570596803?_ga=2.41737122.1878036391.1680456715-375155477.1680456715&_gl=1*1p2e9cf*_ga*MTM3NTE1NTQ3Ny4xNjgwNDU2NzE1*_ga_0HYE8YG0M6*MTY4MDQ1NjcxNi4xLjEuMTY4MDQ1NjgzOS4wLjAuMA. Accessed 2 Apr 2023
6. Abou Neel EA, Bozec L, Perez RA, Kim H-W, Knowles JC (2015) Nanotechnology in dentistry: prevention, diagnosis, and therapy. Int J Nanomed 10:6371–6394. https://doi.org/10.2147/IJN.S86033
7. Bapat RA et al (2019) The use of nanoparticles as biomaterials in dentistry. Drug Discov Today 24(1):85–98. https://doi.org/10.1016/j.drudis.2018.08.012
8. Yudaev P, Chuev V, Klyukin B, Kuskov B, Mezhuev Y, Chistyakov E (2022) Polymeric dental nanomaterials: antimicrobial action. Polymers 14(5). https://doi.org/10.3390/polym14050864
9. Yang J, Zhang Z, Men X, Xu X (2009) Fabrication of stable, transparent and superhydrophobic nanocomposite films with polystyrene functionalized carbon nanotubes. Appl Surf Sci 255(22). https://doi.org/10.1016/j.apsusc.2009.07.010
10. Chen C et al (2014) Antibacterial activity and ion release of bonding agent containing amorphous calcium phosphate nanoparticles. Dent Mater 30(8):891–901. https://doi.org/10.1016/J.DENTAL.2014.05.025
11. Subramani K, Jung RE, Molenberg A, Hammerle CHF (2009) Biofilm on dental implants: a review of the literature. Int J Oral Maxillofac Implants 24(4):616–626
12. Abe Y, Skali-Lami S, Block JC, Francius G (2012) Cohesiveness and hydrodynamic properties of young drinking water biofilms. Water Res 46(4):1155–1166. https://doi.org/10.1016/J.WATRES.2011.12.013
13. Pinzón-Arango PA, Nagarajan R, Camesano TA (2010) Effects of L-alanine and inosine germinants on the elasticity of bacillus anthracis spores. Langmuir 26(9):6535–6541. https://doi.org/10.1021/LA904071Y/ASSET/IMAGES/MEDIUM/LA-2009-04071Y_0006.GIF
14. Zhang T, Chao Y, Shih K, Li XY, Fang HHP (2011) Quantification of the lateral detachment force for bacterial cells using atomic force microscope and centrifugation. Ultramicroscopy 111(2):131–139. https://doi.org/10.1016/J.ULTRAMIC.2010.10.005
15. Yang X et al (2010) Atomic force microscopy investigation of the characteristic effects of silver ions on Escherichia coli and Staphylococcus epidermidis. Talanta 81(4–5):1508–1512. https://doi.org/10.1016/J.TALANTA.2010.02.061

16. McKendry RA (2012) Nanomechanics of superbugs and superdrugs: new frontiers in nanomedicine. Biochem Soc Trans 40(4):603–608. https://doi.org/10.1042/BST20120082
17. Sullan RMA et al (2014) Single-cell force spectroscopy of pili-mediated adhesion. Nanoscale 6(2). https://doi.org/10.1039/c3nr05462d
18. Longo G et al (2012) Force volume and stiffness tomography investigation on the dynamics of stiff material under bacterial membranes. J Mol Recognit 25(5):278–284. https://doi.org/10.1002/JMR.2171
19. Malacarne-Zanon J et al (2010) Permeability of dental adhesives—a SEM assessment. Eur J Dent 4(4):429–439
20. Hashimoto M et al (2004) In vitro effect of nanoleakage expression on resin-dentin bond strengths analyzed by microtensile bond test, SEM/EDX and TEM. Biomaterials 25(25). https://doi.org/10.1016/j.biomaterials.2004.01.009
21. Profeta AC, Mannocci F, Foxton RM, Thompson I, Watson TF, Sauro S (2012) Bioactive effects of a calcium/sodium phosphosilicate on the resin-dentine interface: a microtensile bond strength, scanning electron microscopy, and confocal microscopy study. Eur J Oral Sci 120(4). https://doi.org/10.1111/j.1600-0722.2012.00974.x
22. Morães RR et al (2012) Improved dental adhesive formulations based on reactive nanogel additives. J Dent Res 91(2). https://doi.org/10.1177/0022034511426573
23. Lohbauer U et al (2010) Zirconia nanoparticles prepared by laser vaporization as fillers for dental adhesives. Acta Biomater 6(12). https://doi.org/10.1016/j.actbio.2010.07.002
24. Wagner A, Belli R, Stötzel C, Hilpert A, Müller FA, Lohbauer U (2013) Biomimetically-and hydrothermally-grown HAp nanoparticles as reinforcing fillers for dental adhesives. J Adhes Dent 15(5). https://doi.org/10.3290/j.jad.a29534
25. Di Hipólito V, Reis AF, Mitra SB, de Goes MF (2012) Interaction morphology and bond strength of nanofilled simplified-step adhesives to acid etched dentin. Eur J Dent 6(4):349–360
26. Ouyang X et al (2011) Synthesis and characterization of triethylene glycol dimethacrylate nanocapsules used in a self-healing bonding resin. J Dent 39(12). https://doi.org/10.1016/j.jdent.2011.09.001
27. Yoshihara K et al (2010) Nano-controlled molecular interaction at adhesive interfaces for hard tissue reconstruction. Acta Biomater 6(9). https://doi.org/10.1016/j.actbio.2010.03.024
28. Chung JH et al (2016) Synthesis, characterization, biocompatibility of hydroxyapatite-natural polymers nanocomposites for dentistry applications. Artif Cells Nanomed Biotechnol 44(1). https://doi.org/10.3109/21691401.2014.944644
29. Sharifi S, Khosroshahi AZ, Dizaj SM, Rezaei Y (2022) Preparation, physicochemical assessment and the antimicrobial action of hydroxyapatite–gelatin/curcumin nanofibrous composites as a dental biomaterial. Biomimetics 7(1). https://doi.org/10.3390/biomimetics7010004
30. Abdelshafi MA, Fathy SM, Elkhooly TA, Reicha FM, Osman MF (2021) Bond strength of demineralized dentin after synthesized collagen/hydroxyapatite nanocomposite application. J Mech Behav Biomed Mater 121. https://doi.org/10.1016/j.jmbbm.2021.104590
31. Imani MM, Kiani M, Rezaei F, Souri R, Safaei M (2021) Optimized synthesis of novel hydroxyapatite/CuO/TiO_2 nanocomposite with high antibacterial activity against oral pathogen Streptococcus mutans. Ceram Int 47(23). https://doi.org/10.1016/j.ceramint.2021.08.246
32. Balu SK, Sampath V, Andra S, Alagar S, Manisha Vidyavathy S (2021) Fabrication of carbon and silver nanomaterials incorporated hydroxyapatite nanocomposites: enhanced biological and mechanical performances for biomedical applications. Mater Sci Eng C 128. https://doi.org/10.1016/j.msec.2021.112296
33. Rashahmadi S, Hasanzadeh R, Mosalman S (2017) Improving the mechanical properties of poly methyl methacrylate nanocomposites for dentistry applications reinforced with different nanoparticles. Polym - Plast Technol Eng 56(16). https://doi.org/10.1080/03602559.2017.1289402
34. Mallakpour S, Khani M, Mallakpour F, Fathi M (2018) Preparation and characterization of polyvinylpyrrolidone/L-leucine amino acid–modified montmorillonite/chiral diacid–functionalized Mg-substituted fluorapatite nanocomposites by ultrasonic-assisted rapid process. Polym - Plast Technol Eng 57(1). https://doi.org/10.1080/03602559.2017.1298804

35. Deng F, Fu LH, Ma MG (2015) Microwave-assisted rapid synthesis and characterization of CaF_2 particles-filled cellulose nanocomposites in ionic liquid. Carbohydr Polym 121. https://doi.org/10.1016/j.carbpol.2014.12.033
36. Sözügeçer S, Bayramgil NP (2021) Preparation and characterization of polyacrylic acid-hydroxyapatite nanocomposite by microwave-assisted synthesis method. Heliyon 7(6). https://doi.org/10.1016/j.heliyon.2021.e07226
37. Karthik K, Dhanuskodi S, Gobinath C, Prabukumar S, Sivaramakrishnan S (2018) Multifunctional properties of microwave assisted CdO–NiO–ZnO mixed metal oxide nanocomposite: enhanced photocatalytic and antibacterial activities. J Mater Sci: Mater Electron 29(7). https://doi.org/10.1007/s10854-017-8513-y
38. Raho R, Paladini F, Lombardi FA, Boccarella S, Zunino B, Pollini M (2015) In-situ photo-assisted deposition of silver particles on hydrogel fibers for antibacterial applications. Mater Sci Eng C 55. https://doi.org/10.1016/j.msec.2015.05.050
39. Aseena S, Abraham N, Suresh Babu V (2019) Synthesis of ZnO/CNT nanocomposite by microwave and ultrasound assisted two step process. In: Proceedings—international conference on vision towards emerging trends in communication and networking, ViTECoN 2019. https://doi.org/10.1109/ViTECoN.2019.8899366
40. Liu MH, Chan CH, Ling JH, Chris Wang CR (2007) Filling in dentinal tubules. Nanotechnology 189(47):475104. https://doi.org/10.1088/0957-4484/18/47/475104
41. Fallis AG (2013) Ingle's endodontics 6, vol 53, no 9
42. Arens D, Torabinejad M, Chivian N, Rubinstein R (1998) Practical lessons in endodontic surgery
43. Gartner AH, Dorn SO (1992) Advances in endodontic surgery. Dent Clin N Am 36(2). https://doi.org/10.1016/s0011-8532(22)02501-0
44. Kishen A, Shi Z, Shrestha A, Neoh KG (2008) An investigation on the antibacterial and antibiofilm efficacy of cationic nanoparticulates for root canal disinfection. J Endod 34(12). https://doi.org/10.1016/j.joen.2008.08.035
45. Lu Z, Rong K, Li J, Yang H, Chen R (2013) Size-dependent antibacterial activities of silver nanoparticles against oral anaerobic pathogenic bacteria. J Mater Sci Mater Med 24(6):1465–1471. https://doi.org/10.1007/S10856-013-4894-5
46. Monzavi A, Eshraghi S, Hashemian R, Momen-Heravi F (2015) In vitro and ex vivo antimicrobial efficacy of nano-MgO in the elimination of endodontic pathogens. Clin Oral Investig 19(2):349–356. https://doi.org/10.1007/S00784-014-1253-Y/FIGURES/2
47. Suzuki T et al (2011) Induced migration of dental pulp stem cells for in vivo pulp regeneration. J Dent Res 90(8):1013–1018. https://doi.org/10.1177/0022034511408426/SUPPL_FILE/DS_10.1177_0022034511408426.PDF
48. Rodríguez-Lozano FJ, Guerrero-Gironés J, Pecci-Lloret MR, Pecci-Lloret MP (2022) Stem cells for endodontic regeneration. In: Biomaterials in endodontics, pp. 273–283. https://doi.org/10.1016/B978-0-12-821746-7.00012-7
49. Hanafy AK, Shinaishin SF, Eldeen GN, Aly RM (2018) Nano hydroxyapatite & mineral trioxide aggregate efficiently promote odontogenic differentiation of dental pulp stem cells. Open Access Maced J Med Sci 6(9):1727. https://doi.org/10.3889/OAMJMS.2018.368
50. Li Z, Xie K, Yang S, Yu S, Xiao Y, Zhou Y (2021) Multifunctional Ca-Zn-Si-based micro-nano spheres with anti-infective, anti-inflammatory, and dentin regenerative properties for pulp capping application. J Mater Chem B 9(39). https://doi.org/10.1039/d1tb01517f
51. Sinjari B et al (2019) Curcumin/liposome nanotechnology as delivery platform for anti-inflammatory activities via NFkB/ERK/pERK pathway in human dental pulp treated with 2-HydroxyEthyl MethAcrylate (HEMA). Front Physiol 10:633. https://doi.org/10.3389/FPHYS.2019.00633/BIBTEX
52. Kim JK et al (2010) Differentiating dental pulp cells via RGD-dendrimer conjugates. 89(12):1433–1438. https://doi.org/10.1177/0022034510384870
53. Elgendy AA, Fayyad DM (2017) Cell viability and apoptotic changes of dental pulp stem cells treated with propolis, chitosan, and their nano counterparts. Tanta Dent J 14(4):198. https://doi.org/10.4103/TDJ.TDJ_27_17

54. Tondnevis F, Ketabi M, Fekrazad R, Sadeghi A, Abolhasani MM (2019) Using chitosan besides nano hydroxyapatite and fluorohydroxyapatite boost dental pulp stem cell proliferation. J Biomim, Biomater Biomed Eng 42:39–50. https://doi.org/10.4028/WWW.SCIENTIFIC.NET/JBBBE.42.39
55. Baskar K et al (2022) Eggshell derived nano-hydroxyapatite incorporated carboxymethyl chitosan scaffold for dentine regeneration: a laboratory investigation. Int Endod J 55(1):89–102. https://doi.org/10.1111/IEJ.13644
56. Lee DK et al (2015) Nanodiamond-gutta percha composite biomaterials for root canal therapy. ACS Nano 9(11):11490–11501. https://doi.org/10.1021/ACSNANO.5B05718
57. Vagropoulou G et al (2021) Hybrid chitosan/gelatin/nanohydroxyapatite scaffolds promote odontogenic differentiation of dental pulp stem cells and in vitro biomineralization. Dent Mater 37(1):e23–e36. https://doi.org/10.1016/J.DENTAL.2020.09.021
58. Alipour M et al (2021) The antimicrobial, antioxidative, and anti-inflammatory effects of polycaprolactone/gelatin scaffolds containing chrysin for regenerative endodontic purposes. Stem Cells Int 2021. https://doi.org/10.1155/2021/3828777
59. Alipour M et al (2021) The osteogenic differentiation of human dental pulp stem cells in alginate-gelatin/nano-hydroxyapatite microcapsules. BMC Biotechnol 21(1). https://doi.org/10.1186/S12896-020-00666-3
60. Ilyas K et al (2018) Microwave-assisted synthesis and evaluation of type 1 collagen–apatite composites for dental tissue regeneration. 33(1):103–115. https://doi.org/10.1177/0885328218773220
61. de Stefani A, Bruno G, Preo G, Gracco A (2020) Application of nanotechnology in orthodontic materials: a state-of-the-art review. Dent J 8(4):126. https://doi.org/10.3390/DJ8040126
62. Yin IX, Zhang J, Zhao IS, Mei ML, Li Q, Chu CH (2020) The antibacterial mechanism of silver nanoparticles and its application in dentistry. Int J Nanomed 15. https://doi.org/10.2147/IJN.S246764
63. Bapat RA et al (2018) An overview of application of silver nanoparticles for biomaterials in dentistry. Mater Sci Eng C 91. https://doi.org/10.1016/j.msec.2018.05.069
64. Akarajarasrod P et al (2021) Antibacterial effect of experimental orthodontic adhesives containing gold nanoparticles against Streptococcus mutans and Streptococcus sobrinus. Key Eng Mater 904:301–308. https://doi.org/10.4028/WWW.SCIENTIFIC.NET/KEM.904.301
65. Alam MK, Alsuwailem R, Alfawzan AA (2022) Antibacterial activity and bond strength of silver nanoparticles modified orthodontic bracket adhesive: a systematic review and meta-analysis of in-vitro and in-vivo studies. Int J Adhes Adhes 113:103040. https://doi.org/10.1016/J.IJADHADH.2021.103040
66. Poosti M, Ramazanzadeh B, Zebarjad M, Javadzadeh P, Naderinasab M, Shakeri MT (2013) Shear bond strength and antibacterial effects of orthodontic composite containing TiO_2 nanoparticles. Eur J Orthod 35(5). https://doi.org/10.1093/ejo/cjs073
67. Eshed M, Lellouche J, Matalon S, Gedanken A, Banin E (2012) Sonochemical coatings of ZnO and CuO nanoparticles inhibit streptococcus mutans biofilm formation on teeth model. Langmuir 28(33). https://doi.org/10.1021/la301432a
68. Eleni PN, Katsavou I, Krokida MK, Polyzois GL, Gettleman L (2009) Mechanical behavior of facial prosthetic elastomers after outdoor weathering. Dent Mater 25(12). https://doi.org/10.1016/j.dental.2009.06.018
69. Mehra M, Vahidi F, Berg RW (2014) A complete denture impression technique survey of postdoctoral prosthodontic programs in the United States. J Prosthodont 23(4). https://doi.org/10.1111/jopr.12099
70. Cuy JL, Mann AB, Livi KJ, Teaford MF, Weihs TP (2002) Nanoindentation mapping of the mechanical properties of human molar tooth enamel. Arch Oral Biol 47(4). https://doi.org/10.1016/S0003-9969(02)00006-7
71. Webster TJ, Ejiofor JU (2004) Increased osteoblast adhesion on nanophase metals: Ti, Ti6Al4V, and CoCrMo. Biomaterials 25(19). https://doi.org/10.1016/j.biomaterials.2003.12.002

72. dos Santos JG, Fonseca RG, Adabo GL, dos Santos Cruz CA (2006) Shear bond strength of metal-ceramic repair systems. J Prosthet Dent 96(3). https://doi.org/10.1016/j.prosdent.2006.07.002
73. Van Bael S, Kerckhofs G, Moesen M, Pyka G, Schrooten J, Kruth JP (2011) Micro-CT-based improvement of geometrical and mechanical controllability of selective laser melted Ti6Al4V porous structures. Mater Sci Eng, A 528(24):7423–7431. https://doi.org/10.1016/J.MSEA.2011.06.045
74. Liu Z, Zhang X, Xuan F, Wang Z, Tu S (2013) Effect of laser power on the microstructure and mechanical properties of TiN/Ti3Al composite coatings on Ti6Al4V. Chin J Mech Eng (Engl Ed) 26(4). https://doi.org/10.3901/CJME.2013.04.714
75. Gendreau L, Loewy ZG (2011) Epidemiology and etiology of denture stomatitis. J Prosthodont 20(4). https://doi.org/10.1111/j.1532-849X.2011.00698.x
76. Jindal SK, Sherriff M, Waters MG, Smay JE, Coward TJ (2018) Development of a 3D printable maxillofacial silicone: part II. Optimization of moderator and thixotropic agent. J Prosthet Dent 119(2). https://doi.org/10.1016/j.prosdent.2017.04.028
77. Bellamy K, Limbert G, Waters MG, Middleton J (2003) An elastomeric material for facial prostheses: synthesis, experimental and numerical testing aspects. Biomaterials 24(27). https://doi.org/10.1016/S0142-9612(03)00412-5
78. Jindal SK, Sherriff M, Waters MG, Coward TJ (2016) Development of a 3D printable maxillofacial silicone: part I. Optimization of polydimethylsiloxane chains and cross-linker concentration. J Prosthet Dent 116(4). https://doi.org/10.1016/j.prosdent.2016.02.020
79. Makode S, Singh G, Chanda A (2021) Development of novel anisotropic skin simulants. Phys Scr 96:125019. https://doi.org/10.1088/1402-4896/AC2EFD
80. Singh G, Chanda A (2023) Development and mechanical characterization of artificial surrogates for brain tissues. Biomed Eng Adv 5:100084. https://doi.org/10.1016/J.BEA.2023.100084
81. Gupta V, Singh G, Chanda A (2023) Development of novel hierarchical designs for skin graft simulants with high expansion potential. Biomed Phys Eng Express 9:035024. https://doi.org/10.1088/2057-1976/ACC661
82. Cevik P, Eraslan O (2017) Effects of the addition of titanium dioxide and silaned silica nanoparticles on the mechanical properties of maxillofacial silicones. J Prosthodont 26(7). https://doi.org/10.1111/jopr.12438
83. Zayed SM, Alshimy AM, Fahmy AE (2014) Effect of surface treated silicon dioxide nanoparticles on some mechanical properties of maxillofacial silicone elastomer. Int J Biomater 2014. https://doi.org/10.1155/2014/750398
84. Goiato MC, Haddad MF, Pesqueira AA, Moreno A, dos Santos DM, Bannwart LC (2011) Effect of chemical disinfection and accelerated aging on color stability of maxillofacial silicone with opacifiers. J Prosthodont 20(7). https://doi.org/10.1111/j.1532-849X.2011.00755.x
85. Chaughule RS, Dashaputra R (2020) Advances in dental implantology using nanomaterials and allied technology applications. In: Advances in dental implantology using nanomaterials and allied technology applications, pp. 1–410. https://doi.org/10.1007/978-3-030-52207-0
86. Le Guéhennec L, Soueidan A, Layrolle P, Amouriq Y (2007) Surface treatments of titanium dental implants for rapid osseointegration. Dent Mater 23(7). https://doi.org/10.1016/j.dental.2006.06.025
87. Yao C, Slamovich EB, Webster TJ (2008) Enhanced osteoblast functions on anodized titanium with nanotube-like structures. J Biomed Mater Res A 85(1). https://doi.org/10.1002/jbm.a.31551
88. Park J, Bauer S, Schlegel KA, Neukam FW, Der Von Mark K, Schmuki P (2009) TiO_2 nanotube surfaces: 15 nm—an optimal length scale of surface topography for cell adhesion and differentiation. Small 5(6). https://doi.org/10.1002/smll.200801476
89. Sjöström T, Dalby MJ, Hart A, Tare R, Oreffo ROC, Su B (2009) Fabrication of pillar-like titania nanostructures on titanium and their interactions with human skeletal stem cells. Acta Biomater 5(5). https://doi.org/10.1016/j.actbio.2009.01.007

90. Takeuchi M, Abe Y, Yoshida Y, Nakayama Y, Okazaki M, Akagawa Y (2003) Acid pretreatment of titanium implants. Biomaterials 24(10). https://doi.org/10.1016/S0142-9612(02)00576-8
91. Yamada M et al (2012) Bone integration capability of nanopolymorphic crystalline hydroxyapatite coated on titanium implants. Int J Nanomed 7:859–873. https://doi.org/10.2147/IJN.S28082
92. Krupa D et al (2004) Effect of calcium and phosphorus ion implantation on the corrosion resistance and biocompatibility of titanium. Biomed Mater Eng 14(4):525–536
93. Rani VVD, Manzoor K, Menon D, Selvamurugan N, Nair SV (2009) The design of novel nanostructures on titanium by solution chemistry for an improved osteoblast response. Nanotechnology 20(19). https://doi.org/10.1088/0957-4484/20/19/195101
94. Sugita Y et al (2011) Effects of pico-to-nanometer-thin TiO_2 coating on the biological properties of microroughened titanium. Biomaterials 32(33). https://doi.org/10.1016/j.biomaterials.2011.07.077
95. Kumar S, Vijayalakshmi R (2006) Nanotechnology in dentistry. Indian J Dent Res 17(2):62. https://doi.org/10.4103/0970-9290.29890
96. Hannig M, Kriener L, Hoth-Hannig W, Becker-Willinger C, Schmidt H (2007) Influence of nanocomposite surface coating on biofilm formation in situ. J Nanosci Nanotechnol. https://doi.org/10.1166/jnn.2007.18117
97. Cross K, Huq N, Reynolds E (2007) Casein phosphopeptides in oral health—chemistry and clinical applications. Curr Pharm Des 13(8). https://doi.org/10.2174/138161207780363086
98. Reynolds EC, Cai F, Shen P, Walker GD (2016) Retention in plaque and remineralization of enamel lesions by various forms of calcium in a mouthrinse or sugar-free chewing gum. 82(3):206–211. https://doi.org/10.1177/154405910308200311
99. Gu H et al (2012) Effect of $ZnCl_2$ on plaque growth and biofilm vitality. Arch Oral Biol 57(4). https://doi.org/10.1016/j.archoralbio.2011.10.001
100. Song B, Leff LG (2006) Influence of magnesium ions on biofilm formation by Pseudomonas fluorescens. Microbiol Res 161(4). https://doi.org/10.1016/j.micres.2006.01.004
101. Xu HHK et al (2006) Nano DCPA-whisker composites with high strength and Ca and PO4 release. J Dent Res 85(8). https://doi.org/10.1177/154405910608500807
102. Roveri N et al (2008) Synthetic biomimetic carbonate-hydroxyapatite nanocrystals for enamel remineralization. Adv Mater Res. https://doi.org/10.4028/www.scientific.net/amr.47-50.821
103. He G, Dahl T, Veis A, George A (2003) Nucleation of apatite crystals in vitro by self-assembled dentin matrix protein 1. Nat Mater 2(8). https://doi.org/10.1038/nmat945
104. Ganesh N, Jayakumar R, Koyakutty M, Mony U, Nair SV (2012) Embedded silica nanoparticles in poly(caprolactone) nanofibrous scaffolds enhanced osteogenic potential for bone tissue engineering. Tissue Eng Part A 18(17–18). https://doi.org/10.1089/ten.tea.2012.0167
105. Woo KM, Chen VJ, Ma PX (2003) Nano-fibrous scaffolding architecture selectively enhances protein adsorption contributing to cell attachment. J Biomed Mater Res A 67(2). https://doi.org/10.1002/jbm.a.10098

Chapter 10
Biofidelic Tongue and Tonsils Tissue Surrogates

Gurpreet Singh and Arnab Chanda

1 Introduction

The tongue, a key functional organ of the upper airway that aids in swallowing, speaking, licking, and breathing, is made up of muscles and a rough, thin covering that resembles skin. The tongue helps in grasping and swallowing food and is situated posterior and medial to the teeth [1, 2]. The biomedical modeling of the human tongue has been studied by the researchers [1, 3, 4], however, the mechanical properties have not received much attention. There are some notable studies that reported the elastic modulus of human tongue tissue as 6 kPa [5–7] and 15 kPa [2]. Recently, the indentation method was used to estimate the elastic modulus of the human tongue, which was recorded as 5.52 ± 1.19 kPa [8]. The biomechanical properties of the human tongue can be used for the finite element modeling of the upper airway, which would be beneficial to study the respiratory disorders such as obstructive sleep apnoea (OSA) and for the prediction of various therapy (jaw reconstruction and speech therapy) [9–11].

Similar to the tongue, tonsils are a vital part of our immune system that prevents infection in the throat and lungs by limiting the growth of bacteria and viruses and killing them with antibodies produced by immune cells when we breathe. Tonsils are fleshy lumps and two in number, one on each side of the rear of the mouth. The size of one's tonsils may differ from another's, and any illness may cause them to become inflamed. OSA is a chronic condition due to abnormal upper airway deformation, enlarged tonsils, and other oropharyngeal tissues, such as the tongue, soft palate, and uvula. The knowledge of the mechanical properties of the tonsils would be

G. Singh · A. Chanda (✉)
Centre for Biomedical Engineering, Indian Institute of Technology (IIT), Delhi, India
e-mail: arnab.chanda@cbme.iitd.ac.in

A. Chanda
Department of Biomedical Engineering, All India Institute of Medical Sciences (AIIMS), Delhi, India

A. Chanda et al. (eds.), *Materials for Biomedical Simulation*, Materials Horizons: From Nature to Nanomaterials, https://doi.org/10.1007/978-981-99-5064-5_10

indispensable for developing biomechanical models for various disease scenarios such as upper airway collapse simulations to evaluate the obstructive sleep apnoea conditions [8]. In addition, such FE-based model analysis with realistic mechanical properties would be beneficial for the planning of minimal invasive surgery.

It was observed from the literature that the research on both the tissues (i.e., tongue and tonsils) is limited and only a few studies reported the testing of cadaveric samples. The ethical and biosafety issues make it challenging to handle and test such tissues in an experimental setup. In the present study, the focus was on fabricating the tongue and tonsils tissue surrogates and mimic their mechanical properties using four-part silicone-based elastomer material. The non-linear behavior of the fabricated samples was characterized using hyperelastic curve fit models (Mooney-Rivlin and Neo-Hookean) and their prediction accuracies in hyperelastic models were determined in terms of R^2. Section 2 discusses the fabrication of sample coupons for tongue and tonsils tissues, their mechanical testing, and non-linear characterization using different consecutive hyperelastic models. The detailed experimental results, repeatability tests and the results of curve fit models were discussed in Sect. 3. Section 4 summarized the study with conclusions, limitations, and future recommendations of the presented work.

2 Materials and Methods

2.1 Fabrication of Human Tongue and Tonsils Tissue Surrogates

The sample coupons for the tongue and tonsils tissue surrogates were fabricated using silicone-based elastomer material of different shore hardness (5A, 15A, and 30A). The silicone-based elastomer material was procured from Chemzest Enterprises, Chennai, India. By arbitrarily mixing the material of different shore hardness, 15 sample coupons were fabricated (as per detail given in Table 1) for identifying the sample coupons mimicking the mechanical properties of the human tongue and tonsils tissue surrogates. The sample coupons ranging from shore hardness 5A to 15A were made by mixing and varying the ratio-proportion of silicone-based elastomer material of shore hardness 5A and 15A (shown in Table 1). Similarly, to fabricate the sample coupons ranging from 16 to 30A, the silicone-based elastomer material of shore hardness 15A and 30 was mixed in proportion as mentioned in Table 1. The mixing of two-part of one material with another two-part material of different shore hardness results in the four-part elastomer solution of required shore hardness. The mixture was then poured into the mold (Fig. 1) and left to cure for 5–6 h. The shore durometer hardness scale (ASTM-D2240) was used to confirm the shore hardness of the fabricated sample coupons. From here onwards, the parts A and B of the material with shore hardness 5A will be referred to as 1A and 1B, the parts A and B of the elastomer with a 15A shore hardness referred to as 2A and 2B, whereas the parts A

Table 1 Compositional description of silicone-based elastomer material for fabricating the sample coupons (by wt.%)

Sr. No.	Composition of sample coupon	Shore hardness of sample coupon
Test 1	1A50%-1B50%	5A
Test 2	1A45%-1B45%-2A5%-2B5%	6A
Test 3	1A35%-1B35%-2A15%-2B15%	8A
Test 4	1A25%-1B25%-2A25%-2B25%	10A
Test 5	1A15%-1B15%-2A35%-2B35%	12A
Test 6	1A5%-1B5%-2A45%-2B45%	14A
Test 7	2A50%-2B50%	15A
Test 8	2A47%-2B47%-3A3%-3B3%	16A
Test 9	2A40%-2B40%-3A10%-3B10%	18A
Test 10	2A33%-2B33%-3A17%-3B17%	20A
Test 11	2A27%-2B27%-3A23%-3B23%	22A
Test 12	2A20%-2B20%-3A30%-3B30%	24A
Test 13	2A13%-2B13%-3A37%-3B37%	26A
Test 14	2A7%-2B7%-3A43%-3B43%	28A
Test 15	3A50%-3B50%	30A

and B of the silicone-based elastomer with a 30A shore hardness will be referred to as 3A and 3B.

2.2 Mechanical Testing of Fabricated Sample Coupons

The fabricated sample coupons were tested under uniaxial tensile loading at six different strain rates. Figure 2 showed the setup and an ongoing experiment on a universal testing machine (UTM, Finetechno Engineering Pvt. Ltd., Kolkata, India). The six different strain rates were 0.016, 0.4, 0.5, 0.8, 1, and 2.5 mm/s and were chosen from the literature studies [12–17]. Each experimental run was repeated thrice to ensure the accuracy, and the mean values of the respective trials were used to plot the results. It should be noted that a number of considerations must be taken into account for the testing of human soft tissues or soft materials, such as silicone-based elastomer [18–20]. First, special grips were used to provide the friction against slippage during uniaxial tensile testing. Second, for the precise comparison of results with the literature, certain strain rates must be used since they have a significant impact on the load-deformation outcomes of soft materials during tensile testing [18, 21, 22]. An initial load of 0.1 N was applied to confirm that there should be no slack before the start of the experimental test. The reported work was performed considering all these points and accordingly, the experimental trials were conducted.

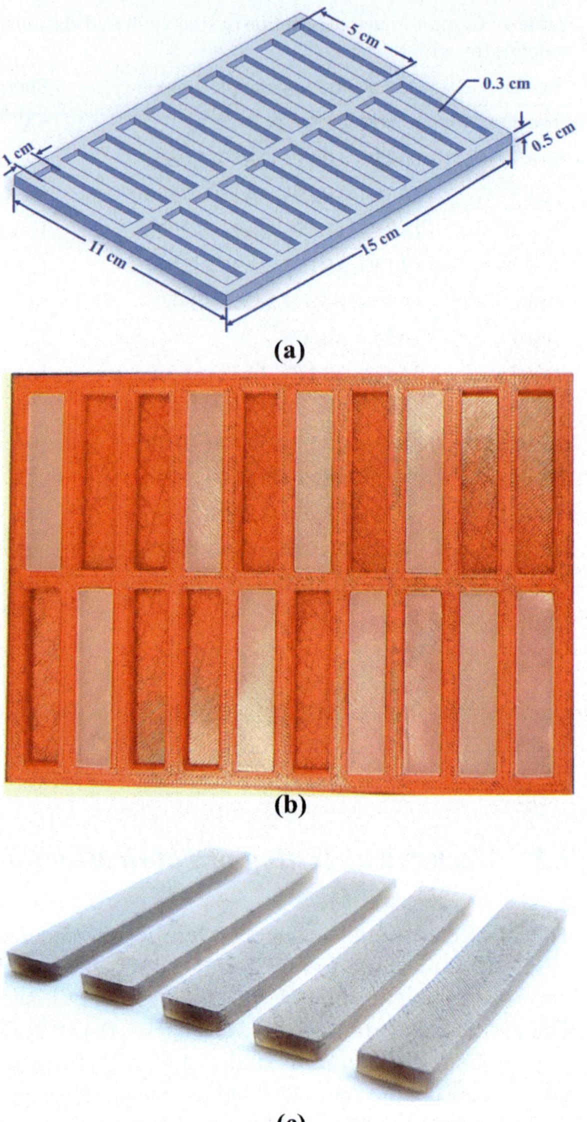

Fig. 1 SolidWorks designed mold in (**a**), mold with sample coupons in (**b**), and fabricated coupons in (**c**)

2.3 Non-linear Material Modeling

Three hyperelastic curve fit models namely, Mooney-Rivlin model, Neo-Hookean and Yeoh model were used to characterize the non-linear behavior of the tested coupons. These material-specific models are based on the strain-energy function (i.e., ψ), and depend either on the invariants of Cauchy-Green tensors (I_1, I_2, and I_3) or the principal stresses (λ_1, λ_2, λ_3) (see Eqs. 1–4) [12, 18, 23–28].

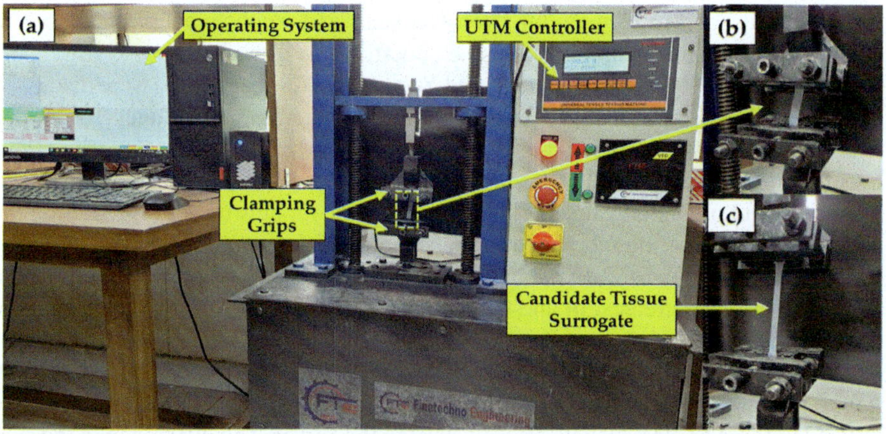

Fig. 2 Setup for UTM testing in (**a**), ongoing tensile test of test coupon in (**b**) and (**c**)

$$\psi = \psi(I_1, I_2, I_3) \tag{1}$$

$$I_1 = \sum_{i=1}^{3} \lambda_i^2 \tag{2}$$

$$I_2 = \sum_{i,j=1}^{3} \lambda_i^2 \lambda_j^2, i \neq j \tag{3}$$

$$I_3 = \prod_{i=1}^{3} \lambda_i^2 \tag{4}$$

The principal Cauchy stress is expressed in the stretch and strain energy function as shown in Eq. 5. Using Eq. 5 for principal Cauchy stress of Mooney-Rivlin, Neo-Hookean and Yeoh models, the non-linear behavior of the uniaxially tested coupons can be estimated using Eqs. 6–8, respectively [29].

$$\sigma_1 = \lambda_1 \frac{\partial \psi}{\partial \lambda_1} - \lambda_3 \frac{\partial \psi}{\partial \lambda_3}, \sigma_2 = \sigma_3 = 0 \tag{5}$$

$$\sigma_{\text{Mooney-Rivlin}} = 2\left(\lambda^2 - \frac{1}{\lambda}\right)\left(c_1 + c_2 \frac{1}{\lambda}\right) \tag{6}$$

$$\sigma_{\text{Neo-Hookean}} = 2\left(\lambda^2 - \frac{1}{\lambda}\right)c_1 \tag{7}$$

$$\sigma_{\text{Yeoh}} = 2\left(\lambda^2 - \frac{1}{\lambda}\right)\left(c_1 + 2c_2(I_1 - 3) + 3c_3(I_1 - 3)^2\right) \tag{8}$$

The results of Mooney-Rivlin, Neo-Hookean and Yeoh hyperelastic curve fit models for the tested sample coupons i.e., coefficients of the Mooney-Rivlin model (c_1 and c_2), Neo-Hookean model (c_1), and Yeoh model (c_1, c_2, and c_3) were discussed in Sect. 3. The Yeoh model is a three-term model whereas the Mooney-Rivlin is a two-term model and Neo-Hookean model is a one-term model. To ensure the precise hyperelastic curve fitting, the R^2 correlation method was used in each case to calculate actual and predicted values. The accuracy of the models used for predicting the material behavior of the fabricated tissue samples was assessed using the R^2 correlation index of the consecutive hyperelastic models. The R^2 value varies from 0 to 1, with 1 being the greatest fit and 0 denoting the poorest fit. Microsoft Excel curve fit solver, which employs a typical generalized reduced gradient (GRG) non-linear optimization method was used to fit the experimental stress–strain data in Eqs. 6–8. Arbitrary hyperelastic equation parameters and stretch values (between 1 and 2) were utilized to forecast stress levels prior to starting the solver. The difference in the sum of squares between the experimental and predicted stress values was calculated for each stretch value. This value was then entered into the Excel curve fit solver together with the selected arbitrary parameters. After solving, the Excel curve fit solver gives the optimum curve fit parameters.

3 Results and Discussion

3.1 Mechanical Behavior of the Tested Samples for Tongue and Tonsils Tissue Surrogates

As mentioned in Sect. 2, the fabricated test coupons of varying shore hardness were tested using uniaxial tensile loading to evaluate their mechanical behavior. The engineering stress versus engineering strain responses for all the 15 experimental tests were compared with the literature stress versus strain bound of the cadaveric tongue tissue and plotted in Fig. 3. Across all the 15 tested samples, the sample coupons of shore hardness 5A, 6A, and 8A showed considerable results and mimic the mechanical properties of the human tongue tissue. The sample coupon of shore hardness 5A exhibits the significant mimicking of mechanical properties and was below the literature bound upto the strain of 1.6. On the other hand, the sample coupons of shore hardness 6A and 8A closely mimic the mechanical behavior of the tongue tissue surrogate (see Fig. 4).

Similar to the tongue tissue surrogates, the experimental results were compared with the literature to identify the test coupon mimicking the mechanical properties of the tonsils tissue. Figure 5 shows the engineering stress versus engineering strain responses of the experimental tests and the literature bound for the cadaveric tonsils tissue. Across all the 15 tested samples, the sample coupon of shore hardness 5A was quantified as the surrogate of tonsils tissue which mimics the mechanical properties of the human tonsils tissue (see Fig. 6). It should be noted that all the engineering

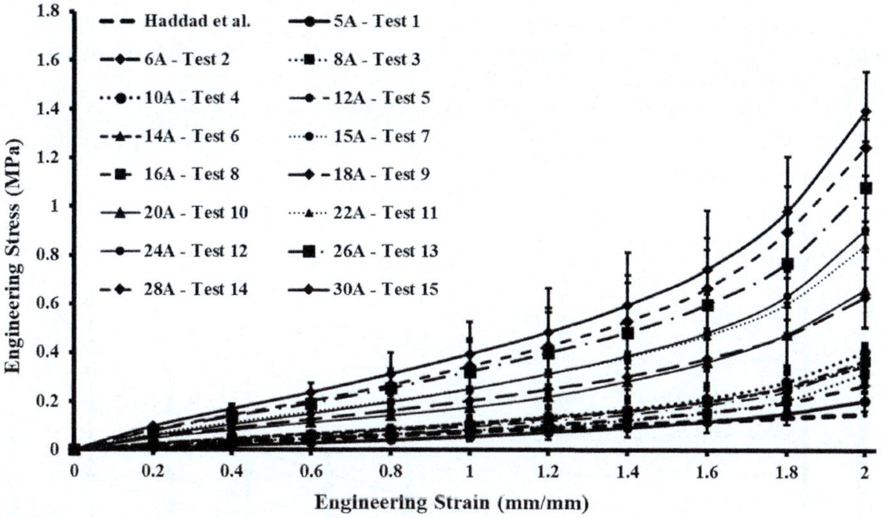

Fig. 3 Stress versus strain plot of the candidate tongue tissue surrogates and comparison with the literature [8]

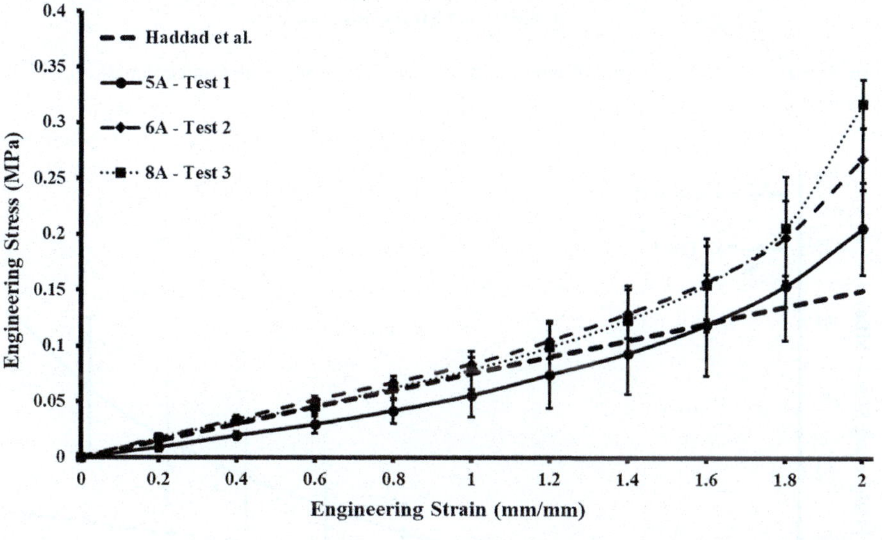

Fig. 4 Stress versus strain plot of the test coupons mimicking the mechanical behavior of tongue tissue [8]

stress versus engineering strain plots showed the average stress values of each tested sample with error bars of standard deviation to indicate the variation of stresses across all strain rates.

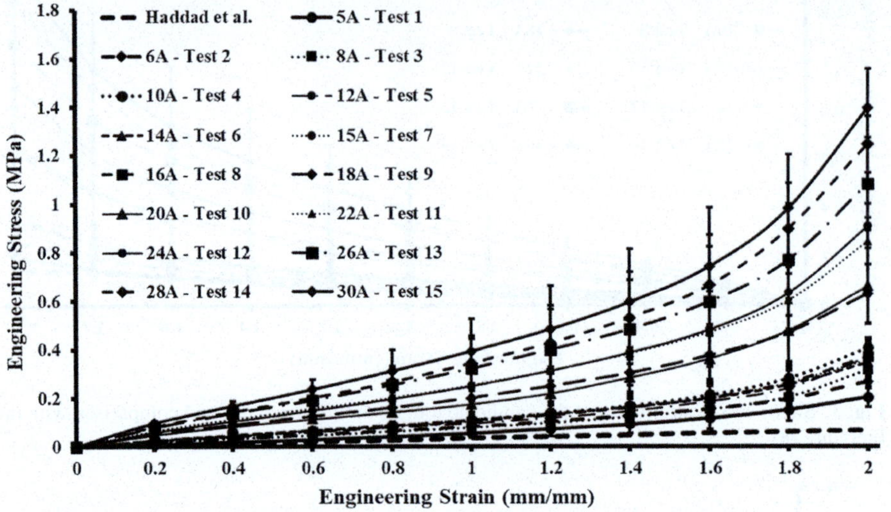

Fig. 5 Stress versus strain plot of the candidate tonsils tissue surrogates and comparison with the literature [8]

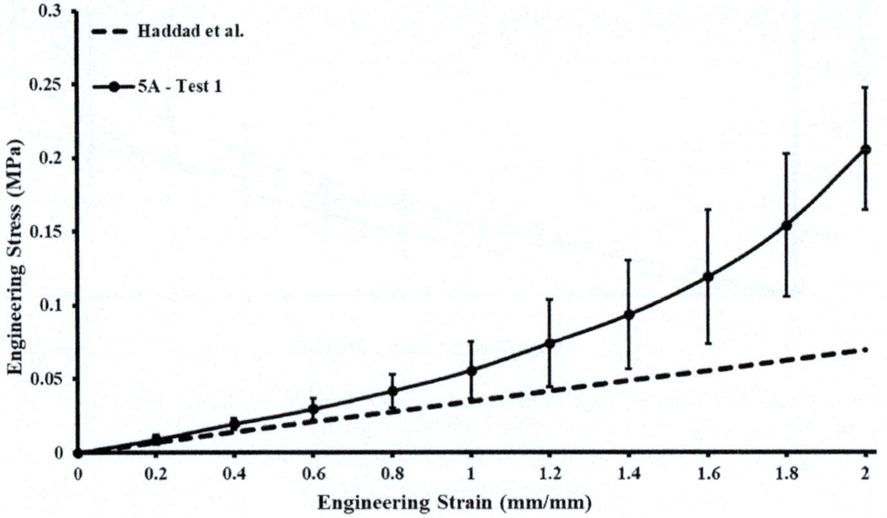

Fig. 6 Stress versus strain plot of the test coupons mimicking the mechanical behavior of tonsils tissue [8]

Table 2 Estimated hyperelastic coefficients of test coupons for tongue and tonsils tissue

Hyperelastic curve fitting results						
Test coupon	Yeoh			Neo-Hookean	Mooney-Rivlin	
	c_1 (MPa)	c_2 (MPa)	c_3 (MPa)	c_1 (MPa)	c_1 (MPa)	c_2 (MPa)
Test 1	4E−03	6.00E−04	1.00E−05	8.76E−03	9.78E−02	7E−04
Test 2	9.61E−03	1.00E−04	1.70E−05	1.24E−02	1.32E−02	1.2E−03
Test 3	9.9E−03	1.20E−04	2.90E−05	1.35E−02	1.41E−02	1.4E−03
Test 4	1.1E−02	9.00E−04	3.50E−05	1.77E−02	1.65E−02	1.5E−03
Test 5	1.25E−02	8.00E−06	4.00E−05	1.26E−02	1.72E−02	2.1E−03
Test 6	1.32E−02	1.00E−05	4.80E−05	1.53E−02	1.75E−02	3.1E−03
Test 7	1.38E−02	1.80E−05	5.60E−05	1.63E−02	1.77E−02	3.6E−03
Test 8	1.39E−02	2.10E−05	6.50E−05	1.73E−02	1.91E−02	3.9E−03
Test 9	2.27E−02	2.60E−05	8.40E−05	2.82E−02	2.95E−02	4.8E−03
Test 10	2.59E−02	3.30E−05	9.00E−05	3.07E−02	3.31E−02	5.9E−03
Test 11	2.87E−02	3.70E−05	1.25E−04	3.90E−02	3.86E−02	6.4E−03
Test 12	3.19E−02	3.90E−05	1.40E−04	4.03E−02	4.31E−02	7.2E−03
Test 13	4.13E−02	4.40E−05	1.41E−04	4.72E−02	5.26E−02	8.5E−03
Test 14	4.35E−02	5.40E−05	1.98E−04	5.57E−02	5.93E−02	9E−03
Test 15	4.96E−02	6.70E−05	2.11E−04	6.07E−02	6.64E−02	1.13E−02

3.2 Non-linear Hyperelastic Curve Fitting

There are different hyperelastic curve fit models which can be used to quantify the non-linear mechanical behavior of soft materials like polymers and human soft tissues [24]. In this study, the non-linear behavior of the developed tissue surrogates was assessed using Yeoh, Mooney-Rivlin and Neo-Hookean curve fit models. For non-linear hyperelastic curve fitting, Eqs. 6–8 were used to curve fit the engineering stress vs engineering strain plots of the experimental tests. The estimated values of c_1, c_2, and c_3 (Yeoh model), c_1 and c_2 (Mooney-Rivlin model) and c_1 (Neo-Hookean model) for the experimentally tested coupons are listed in Table 2. Across all the three hyperelastic models, the average correlation index (R^2) was estimated to be above 0.9 and evaluated as 0.99 for Yeoh model, 0.98 for Neo-Hookean model and 0.95 for the Mooney-Rivlin model.

4 Conclusions

In this work, a four-part elastomer material system was designed to simulate the mechanical properties of the tongue and tonsils tissue. The compositions of the silicone-based elastomer material were varied to fabricate a total of 15 sample coupons, and tested uniaxially at different strain rates to quantify the compositions mimicking the mechanical properties of tongue and tonsils tissue. Based on the testing results, the mechanical behavior of the tested samples was evaluated and the compositions mimicking properties of tongue and tonsils tissue, respectively were determined and compared with the literature. The hyperelastic curve fit models were used to estimate the non-linear behavior of the fabricated test samples for tongue and tonsils tissues. The development of tissue surrogates for tongue and tonsils tissues would be indispensable for the modeling of a wide range of injury, disease and biomechanics scenarios (such as obstructive sleep apnoea).

The four-part elastomer material system used in the present work has some limitations, which should be acknowledged. In this work, the developed tissue surrogates were tested at six varying strain rates. A wide range of strain rates can be considered in future work to deep insight into the topic. Also, multi-axial loading conditions can be considered to investigate the significance of stresses developed in multiple directions. To mimic the mechanical properties of the tongue and tonsils tissues more effectively, these limitations can be considered and incorporated into future studies.

Acknowledgements Gurpreet Singh is grateful to the Ministry of Education, Government of India, for awarding the Prime Minister's Research Fellowship (Ref: IITD/Admission/Ph.D./PMRF/2020-21/4062) for pursuing his doctoral research program at IIT-Delhi, India.

Conflict of Interest Statement The authors declare no conflict of interest with respect to the research, authorship, and/or publication of this article.

References

1. Kajee Y, Pelteret J-PV, Reddy BD (2013) The biomechanics of the human tongue. Int J Numer Methods Biomed Eng 29:492–514. https://doi.org/10.1002/cnm.2531
2. Payan Y, Bettega G, Raphaël B (1998) A biomechanical model of the human tongue and its clinical implications. In: Lecture notes in computer science (including subseries lecture notes in artificial intelligence and lecture notes in bioinformatics), vol 1496. Springer Verlag, pp 688–695. https://doi.org/10.1007/bfb0056255
3. Gérard JM, Ohayon J, Luboz V, Perrier P, Payan Y (2004) Indentation for estimating the human tongue soft tissues constitutive law: application to a 3D biomechanical model. In: Lecture notes in computer science (including subseries lecture notes in artificial intelligence and lecture notes in bioinformatics), vol 3078, pp 77–83. https://doi.org/10.1007/978-3-540-25968-8_9
4. Hermant N, Perrier P, Payan Y (2017) Human tongue biomechanical modeling. In: Biomechanics of living organs: hyperelastic constitutive laws for finite element modelling. Elsevier Inc., pp 395–411. https://doi.org/10.1016/B978-0-12-804009-6.00019-5
5. Malhotra A, Huang Y, Fogel RB, Pillar G, Edwards JK, Kikinis R et al (2002) The male predisposition to pharyngeal collapse: importance of airway length. Am J Respir Crit Care Med 166:1388–1395. https://doi.org/10.1164/rccm.2112072
6. Huang Y, White DP, Malhotra A (2007) Use of computational modeling to predict responses to upper airway surgery in obstructive sleep apnea. Laryngoscope 117:648–653. https://doi.org/10.1097/MLG.0b013e318030ca55

7. Xu C, Brennick MJ, Dougherty L, Wootton DM (2009) Modeling upper airway collapse by a finite element model with regional tissue properties. Med Eng Phys 31:1343–1348. https://doi.org/10.1016/j.medengphy.2009.08.006
8. Haddad SMH, Dhaliwal SS, Rotenberg BW, Samani A, Ladak HM (2018) Estimation of the Young's moduli of fresh human oropharyngeal soft tissues using indentation testing. J Mech Behav Biomed Mater 86:352–358. https://doi.org/10.1016/j.jmbbm.2018.07.004
9. Stavness I, Hannam AG, Lloyd JE, Fels S (2008) Towards predicting biomechanical consequences of jaw reconstruction. In: Proceedings of the 30th annual international conference of the IEEE engineering in medicine and biology society, EMBS'08—"personalized healthcare through technology". IEEE Computer Society, pp 4567–4570. https://doi.org/10.1109/iembs.2008.4650229
10. Gerard JM, Ohayon J, Luboz V, Perrier P, Payan Y (2005) Non-linear elastic properties of the lingual and facial tissues assessed by indentation technique: application to the biomechanics of speech production. Med Eng Phys 27:884–892. https://doi.org/10.1016/j.medengphy.2005.08.001
11. Buchaillard S, Brix M, Perrier P, Payan Y (2007) Simulations of the consequences of tongue surgery on tongue mobility: implications for speech production in post-surgery conditions. Int J Med Robot Comput Assist Surg 3:252–261. https://doi.org/10.1002/rcs.142
12. Makode S, Singh G, Chanda A (2021) Development of novel anisotropic skin simulants. Phys Scr 96:125019. https://doi.org/10.1088/1402-4896/AC2EFD
13. Singh G, Chanda A (2023) Development and mechanical characterization of artificial surrogates for brain tissues. Biomed Eng Adv 5:100084. https://doi.org/10.1016/J.BEA.2023.100084
14. Gupta V, Singh G, Chanda A (2023) Development of novel hierarchical designs for skin graft simulants with high expansion potential. Biomed Phys Eng Express 9:035024. https://doi.org/10.1088/2057-1976/ACC661
15. Gupta V, Singh G, Chanda A (2023) High expansion auxetic skin graft simulants for severe burn injury mitigation. Eur Burn J 4:108–120. https://doi.org/10.3390/EBJ4010011
16. Singh G, Chanda A (2023) Development and biomechanical testing of artificial surrogates for vaginal tissue. Adv Mater Process Technol. https://doi.org/10.1080/2374068X.2023.2198837
17. Singh G, Chanda A (2023) Biofidelic gallbladder tissue surrogates. Adv Mater Process Technol. https://doi.org/10.1080/2374068X.2023.2198835
18. Chanda A (2018) Biomechanical modeling of human skin tissue surrogates. Biomimetics 3:18. https://doi.org/10.3390/BIOMIMETICS3030018
19. Singh G, Chanda A (2021) Mechanical properties of whole-body soft human tissues: a review. Biomed Mater 16:062004. https://doi.org/10.1088/1748-605X/AC2B7A
20. Chanda A, Unnikrishnan V, Lackey K, Robbins J (2019) Biofidelic conductive soft tissue surrogates. 69:127–135. https://doi.org/10.1080/00914037.2018.1552856
21. Singh G, Gupta V, Chanda A (2022) Artificial skin with varying biomechanical properties. Mater Today Proc. https://doi.org/10.1016/J.MATPR.2022.03.433
22. Chanda A, Callaway C (2018) Tissue anisotropy modeling using soft composite materials. Appl Bionics Biomech 2018. https://doi.org/10.1155/2018/4838157
23. Chanda A, Callaway C, Clifton C, Unnikrishnan V (2016) Biofidelic human brain tissue surrogates. Mech Adv Mater Struct 25:1335–1341. https://doi.org/10.1080/15376494.2016.1143749
24. Chanda A, Graeter R (2018) Human skin-like composite materials for blast induced injury mitigation. J Compos Sci 2:44. https://doi.org/10.3390/jcs2030044
25. Chanda A, Unnikrishnan V, Richter HE, Lockhart ME (2016) A biofidelic computational model of the female pelvic system to understand effect of bladder fill and progressive vaginal tissue stiffening due to prolapse on anterior vaginal wall. Int J Numer Methods Biomed Eng 32:e02767. https://doi.org/10.1002/cnm.2767
26. Chanda A, Curry K (2018) Patient-specific biofidelic human coronary artery surrogates. J Mech Med Biol 18. https://doi.org/10.1142/S0219519418500495
27. Chanda A, Unnikrishnan V, Roy S, Richter HE (2015) Computational modeling of the female pelvic support structures and organs to understand the mechanism of pelvic organ prolapse: a review. Appl Mech Rev 67. https://doi.org/10.1115/1.4030967/370016

28. Chanda A, Chatterjee S, Gupta V (2020) Soft composite based hyperelastic model for anisotropic tissue characterization. J Compos Mater 54:4525–4534. https://doi.org/10.1177/0021998320935560
29. Martins PALS, Jorge RMN, Ferreira AJM (2006) A comparative study of several material models for prediction of hyperelastic properties: application to silicone-rubber and soft tissues. Strain 42:135–147. https://doi.org/10.1111/j.1475-1305.2006.00257.x

Chapter 11
Polymeric Biomaterials for Bioprinting Applications

Akhil Kumar Sonkar, Abhishek Kundu, Deepmala Sharma, Vishnu Agarwal, and Arnab Chanda

1 Introduction

Tissue engineering, regenerative medicine and resettlement are being revolutionised by 3D printing, this allows for the development of very sophisticated, modular, and flexible prosthetics, orthoses, and scaffolds that are patient-specific. For the purpose of tissue engineering, regenerative medicine, and the rehabilitation of patients with severe neurological illnesses, three-dimensional printing (3DP) is becoming a burgeoning technology (such as amyotrophic lateral sclerosis (ALS), craniocerebral trauma, and traumatic spinal cord injury). This is because 3D printing can offer very complex structural designs, tolerant-specific designs, and quick on-demand manufacture at a cheap cost [1–3]. Bioink, the key component of 3D bioprinting, is essential for creating functioning organs or tissue structures. There are several crucial characteristics that must be considered while choosing the bioinks for 3D printing. To create more efficient bioinks for the 3D printing (3DP) of biological or tissue structures, a variety of techniques and property improvements are needed [4]. Materials for 3D printing (3DP) are selected in accordance with the intended use. Materials suitable for 3DP processes include composites, metals, ceramics, and polymers. As polymers

A. K. Sonkar · A. Kundu (✉)
Department of Applied Mechanics, Motilal Nehru National Institute of Technology Allahabad, Prayagraj 211004, India
e-mail: abhishekkunduamd@mnnit.ac.in

D. Sharma
Department of Mathematics, National Institute of Technology Raipur, Chhattisgarh 492010, India

V. Agarwal
Department of Biotechnology, Motilal Nehru National Institute of Technology Allahabad, Prayagraj 211004, India

A. Chanda
Centre for Biomedical Engineering, Indian Institute of Technology (IIT) Delhi, Delhi 110016, India

A. Chanda et al. (eds.), *Materials for Biomedical Simulation*, Materials Horizons: From Nature to Nanomaterials, https://doi.org/10.1007/978-981-99-5064-5_11

are typically inexpensive and readily available, they are one of the most regularly used techniques for 3D manifestation, prototyping, validation, and practical applications. One example is the polycaprolactone (PCL) based printing of tissue scaffolds [5, 6]. In this regard, choosing the right biopolymer to use with bioprinting technology is a major difficulty because doing so leaves cytotoxicity and compatibility problems unstudied. In this chapter, we examine novel biopolymers like chitosan, collagen, hyaluronic acid, and silk fibroin. When choosing a biopolymer, it is vital to resemble the biopolymer closer toward nature. For usage in tissue engineering and reincarnate medicament applications, several various Bioink formulations, for example, cell-biomaterials-based Bioinks and cell-based Bioinks, such as cell aggregates and tissue spheroids, have been characterised. The utilisation of ramification polymeric biomaterials, their alterations, and the incorporation of cells and hydrogels have led to the creation of more adaptive bioinks that are printable, mechanically stable after printing, and compatible with live cells [7]. Additionally, biomaterial inks must be readily fabricated, processed, reasonably priced, and commercially available. Keep in mind that the printing process and the intended use of such materials will determine how they must adhere to each of these standards. Thus, a balance would need to be established between all of these variables in order to develop acceptable printed biomaterials with robust properties for a variety of biomedical application demands [8, 9]. The complex geometrical data received from medical imaging techniques like as X-radiation image, nuclear magnetic resonance imaging (NMRI), and micro-computerised tomography scan (μCT scan) may be used to create and evolve such complex 3D tissue architectures in computer-aided design. The ability to create personalised designs for specific patients, high levels of precision, low costs, and the quick manufacturing of complex structures on-demand are all well-being of 3D bioprinting in biomedical engineering [9, 10]. Due to their capacity to mimic the cellular environment and the wide range of processing choices available for polymeric systems, polymers have made up the great majority of the materials utilised in bioprinting under ambient or only moderately harsh chemical and environmental conditions. Additionally, included are applications for bioprinting, cartilaginous bioplants, wound healing bio-bandages, and tissue engineering solutions. Additionally, it is applied in pharmaceutical manufacturing, cancer research, and tissue transplantation via bioprinting. The natural polymers used as inks in bioprinting are covered in great depth in the current study. The present study evaluates the benefits and drawbacks of recently developed polymeric biomaterials with 3D bioprinting technology [11]. Figure 1 shows the year-wise milestones of 3D printing for tissue engineering.

1.1 Biomaterials and Its Classification

A biomaterial is any material, natural or artificial, that is all or part of a living thing or a biomedical device and that carries out, enhances, or replaces a biological function. The term "biomaterials" refers to a broad variety of tissue engineering techniques that allow the evolution of materials for the repair or replacement of different living system

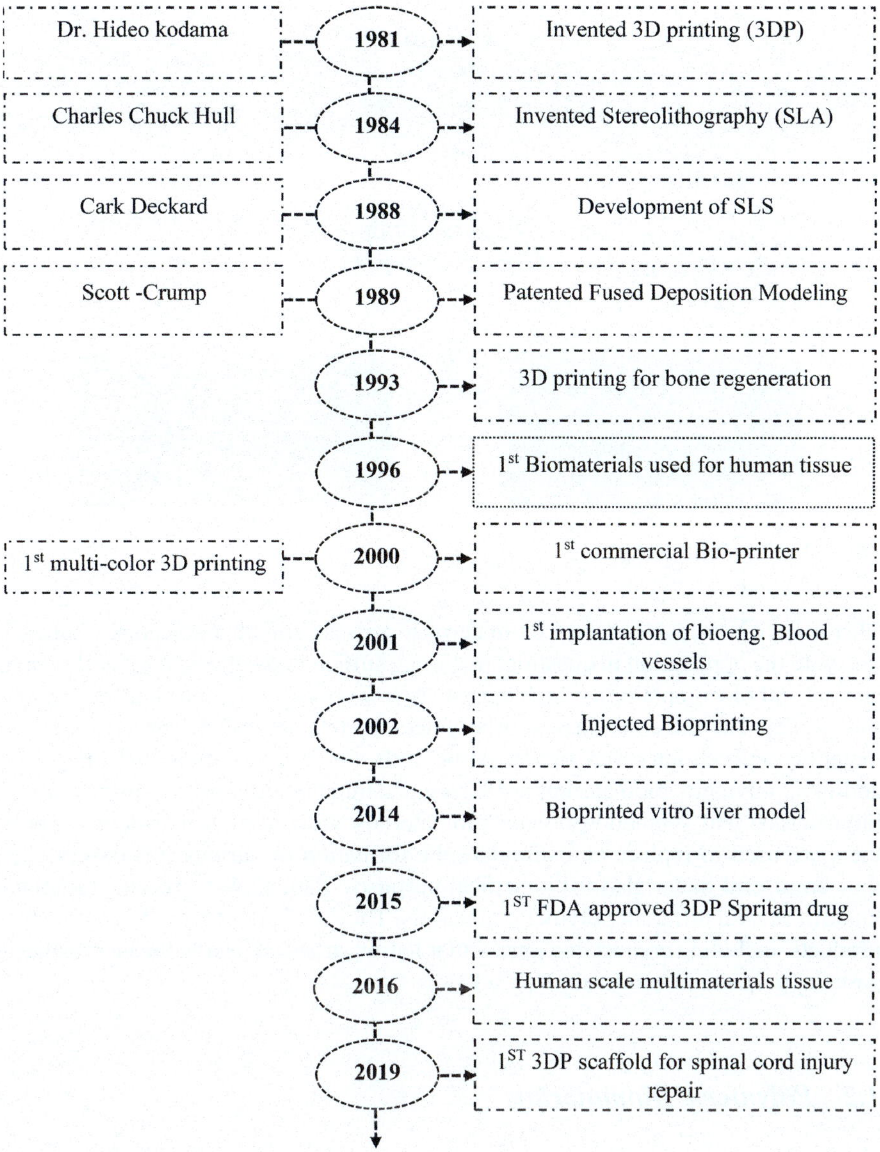

Fig. 1 3D printing milestones for tissue engineering

components. Given that older people are more likely to experience hard tissue failure, demand has increased fast in recent decades as a result of the global phenomenon of population ageing. The main objective is to prolong the life of the implanted biomaterial or to guarantee that it does so until the end of life without faults or the need for revision surgery, contributing to an improvement in the quality of life for the patient.

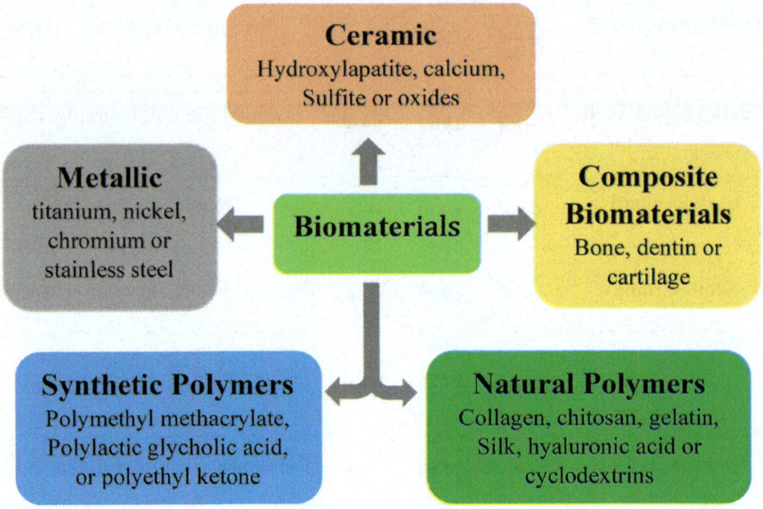

Fig. 2 Types of biomaterials

Therefore, it is necessary for the biomaterial to possess a number of features that are in line with the application in question, including sufficient mechanical reliability, high abrosia and biocompatibility, high wear-tear, low abrasion, and mechanical compatibility [12]. The material classes of polymers, metals, natural materials, composite materials, and ceramics were used to particularly create these biomaterials (Fig. 2). In addition, polymeric biomaterials are segregated into two categories: natural polymer biomaterials and synthetic polymer biomaterials. In cardiovascular devices, polymers are used to replace and stimulate the formation of various soft tissues since they are useful materials for biomedical purposes. Patients have received implants made of several different polymeric materials. They are currently used in a variety of products, including contact and intraocular lenses, artificial hearts, breast prosthesis, vascular grafts, and heart valves [13, 14].

1.2 Polymeric Biomaterials

Polymeric biomaterials have various advantages done other types of materials, including convenience of fabrication, ease clarity of secondary processing, utility with specific mechanical and physical properties, and reasonable pricing. Synthetic and natural polymers are the two kinds employed in biomedical applications. Examples include synthetic polymeric systems including acrylics, polyamides, polyesters, polyethylene, polysiloxanes, and polyurethane. Although synthetic polymers are easily processed, their main drawback is that they are often not biocompatible, which means that using them is frequently accompanied by inflammatory responses. Utilising natural polymers will help solve this issue. For instance, the biomedical industry uses natural polymers like chitosan, carrageenan, and alginate for tissue regeneration and drug delivery systems [15]. The main types of biomaterials used in tissue engineering include artificial polymers, which include hydrophobic compounds like hydroxy acids, poly-lactic-co-glycolic acid (PLGA), and polyanhydrides as well as naturally occurring polymers including complex sugars (hyaluronan, chitosan), inorganics and polysaccharides. There are further structural or functional subgroups, such as those determined by whether they are hydrogels, injectable, surface-modified, capable of delivering pharmaceuticals, are meant for a particular purpose, etc. [16]. The vast range of potential applications for natural polymers based on proteins or polysaccharides in biomedicine has recently received growing interest. These materials are ideal for a number of purposes in industries including tissue engineering, drug delivery, and wound healing due to their chemical substantiality, structural plasticity, biocompatibility, and excellent availability. Biomaterials that have been taken from plants or animals have also undergone modifications to enhance their structural characteristics or promote interactions with adjacent cells and tissues in order to enhance in vivo performance. This has led to creative uses for surface coatings, controlled medication release, and implanted devices [17]. A growing number of polymeric biomaterials are being produced synthetically and naturally, with a variety of uses including tissue engineering, regeneration, and specialised sectors including medication delivery, nanotechnology, and gene therapy [18]. Because of their unique properties, silks are being used in pre-clinical trials and proof-of-concept experiments in sectors including gene therapy and wound healing, in addition to their traditional usage in textiles. By layer-by-layer or spin-coating techniques, silk films are easily made from fibroin stock solutions and have been used as scaffolds in a variety of tissue engineering applications. When osteoblasts were seeded onto scaffolds using functionalised silk films created by chemically combining Arginylglycylaspartic acid (RGD) domains, bone formation was stimulated [19]. Polymer-based biomaterials have taken the place of other materials including metals, alloys, and ceramics due to their low cost, chemical stability, simplicity of processing and reprocessing, and improved corrosion resistance. The medical, biotechnology, food, and cosmetic sectors have all made substantial use of polymer-based biomaterials. Polymeric biomaterials are utilised in medicine for a variety of purposes, including vascular bypass, implantation, wound curing, stitching, catheters, meshes, stents, ligament and tendon renovation, and cardiac procedure valves. Polymeric materials are often categorised as synthetic, natural, or a mix of the two types of polymers for use in biomedical applications. Natural polymers may be derived from both plant and animal sources, with proteins, cellulose, deoxyribonucleic acid (DNA), ribonucleic acid (RNA), cellulose, and silk being the most popular types [20].

1.2.1 Slik Fibroin

Bombyx mori, a silkworm, is the source of the protein fibre known as silk. Spinned by insects and spiders, it is a durable, all-natural biomaterial. Silk fibroin, a protein polymer utilised and investigated for biomaterial applications, is generated from Bombyx mori cocoons. Silk fibroin demonstrates biocompatibility, degrades predictably over time, from hours to years, and is chemically modifiable to modify surface properties or immobilise growth inputs. When moulded into different materials, silk fibroin also possesses exceptional mechanical qualities. Silk biomaterials for various applications can be produced by a number of techniques including the processing of aqueous or organic solvents. Using the techniques outlined in this procedure, silk from Bombyx mori cocoons may be extracted and used to create hydrogels, tubes, sponges, composite materials, fibres, microspheres, and thin films. These materials can be utilised for implantable biomaterials, drug delivery, tissue engineering scaffolds, and in vitro disease modelling [21]. Bombyx mori (silkworm) silk is a special substance that has long been prized for its tenacity and brightness. From ancient times, clinicians have used silk as a suture material. Recently, silk as a biomaterial paid more attention due to a variety of compelling features [22, 23]. These qualities specifically include its biocompatibility, ease of chemical modification, slow rate of in vivo degradation, and capacity for processing into a variety of material forms from either an organic solvent or an aqueous solution [24]. Silk worm silk is more durable than often used polymeric biomaterials like collagen and poly (lactic acid) (PLA). Silk fibres from B. mori have a maximum tensile strength of 740 MPa. Compared to PLA's 28–50 MPa, collagen's 0.9–7.4 MPa ultimate tensile strength is significantly lower. As a result, silk fibroin is an excellent choice of polymer for biological applications. Silk fibres are more resilient compared to the sturdiest synthetic materials [25, 26] (Fig. 3).

(a) Bombyx mori silk cocoon (b) Silk Fiber (c) Silk solution

Fig. 3 Silk fibroin extraction procedure

1.2.2 Collagen

Almost all biological systems include a significant amount of collagen, a crucial protein for all living things. It is a protein that gives bodily tissues rigidity and strength while promoting the development of networks along cellular structures. Because of their great biocompatibility, exceptional biodegradability, and minimal immunogenicity, collagen molecules were shown to be the best candidates for real therapeutic alternatives in contemporary biomedical and biotechnological applications. These factors make collagen one of the biopolymers that are now being researched the most in the medical industry [27]. In both humans and animals, collagen makes up the majority of the structural proteins. It is an important part of the ECM and makes up around 30% of all mammalian proteins. It supports the structural integrity of both hard and soft tissues, such as blood vessels, cartilage, tendon, bone, and ligaments, thanks to its distinctive fibrillar structure [28]. Collagen is only marginally immunogenic, with fibrillose collagen being less immunogenic than smaller molecules as a result of the encapsulation of possible antigenic sites during its auto-polymerisation. In biomedical applications, cross-linking improves a material's biocompatibility, degradability, and resistance to collagenase activity [29]. In order to strengthen its mechanical strength and resistance to enzymatic degradation, extracted collagen must be cross-linked in contrast to decellularised collagen, which has been naturally polymerised in vivo. Cross-linking can be done in a number of well-known ways, including physically using UV or heat treatment, chemically using formaldehyde and glutaraldehyde, or biologically using cross-linking enzymes like transglutaminase [30]. In order to create fibres that mirror the biological properties of ECM at submicron to nanoscale sizes, collagen kinds I, II and III were electrospinning. The collagen fibres were developed with a 67-nm binding pattern that is identical to genuine collagen in order to produce collagen type electrospun biomimetic scaffolds with melodious porosity, mechanical strength, and fibre alignment to topographically direct tissue growth. Electrospun collagen scaffolds have also demonstrated the ability to sustain cellular development [31, 32]. A hydrogel is a 3D polymer network with a large fluid-holding capacity. Due to the hydrophilic functional groups on their polymeric backbone, polymeric hydrogels' major therapeutically significant properties are their capacity to hold water, resistance to disintegration via cross-linking, and flexibility equivalent to that of natural ECM. Because collagen hydrogels have a structural similarity to tissue, they are being researched as biomimicry 3D scaffolds to aid in cell progress [33, 34]. The preservation of cells and bioactive substances by collagen hydrogels makes them desirable scaffolds for tissue engineering. They are essential for the growth of bone and cartilage, functioning as carriers for chondrocytes from cows to offer, among other things, structural integrity and pain-free cartilage articulation [35]. Due in large part to their great biocompatibility, minimal immunogenicity, and diversity in terms of structure, collagen-based materials are now at the forefront of biomaterials used in tissue engineering and regenerative medicament. Applications for collagen in areas including tissue regeneration, medication administration, and wound healing are becoming more and more varied because of improvements in extraction and scaffold composition. In order to increase the in vivo effectiveness of collagen-based scaffolds, future research is anticipated to concentrate on enhancing the mechanical strength, drug transport capacities, and biodegradability of these materials.

1.2.3 Gelatin

Biopolymer made of gelatin is entirely absorbed, biodegradable, and biocompatible. Gelatin is suited for a variety of tissue engineering applications because of these qualities, which have generated a lot of attention. Gelatin is a crucial functional biopolymer that is frequently added to meals to increase their elasticity, consistency, and stability. Together with the skin and bones of land animals, it may also be obtained from fish and insects. To make type A and type B gelatins, respectively, the acid and alkaline procedures are often utilised in gelatin production [36]. The process of thermally denaturing collagen, the most prevalent and essential protein in the animal kingdom, yields gelatin, a dietary item that is essentially entirely composed of protein [37]. High molecular weight gelatin is a necessary hydrocolloid with a polypeptide chain. Because of its capacity to thicken and gel, it has grown in favour among the general public and is used in a range of culinary products. In contrast to other hydrocolloids, which are typically polysaccharides, gelatin is a digestible protein that contains all of the necessary amino acids minus tryptophan. Because of exposure to many environmental conditions, most notably temperature, the amount of amino acids can differ between species, especially for proline and hydroxyproline [38]. From a variety of collagen sources, gelatin may be produced. The main commercial sources are fish, cow bones, pig skins, and pig hides. It might thus originate from both agricultural and non-agricultural sources. Seaweed extracts and other substances referred to as "vegetable gelatin" have no biological connection to gelatin, and gelatin has no plant origins [39]. The name "gelatin" refers to a group of dietary protein products made by hydrolysing collagen from cold-blooded animals like fish and insects, which is found in their bones and skins. There are two methods for extracting the fish and insect gelatin: an acid procedure and an alkaline technique. When discussing the extraction of fish gelatin, the term "acid process" refers to both the extraction carried out in an acid medium and, in certain situations, the use of an acid pre-treatment before the extraction. The extraction can be done in an alkaline, neutral, or acid medium thanks to the alkaline procedure, which involves pre-treating fish skin with an alkaline solution and often neutralising it with an acid solution [40, 41]. The chemical attributes of gelatin are connected to its functional qualities. The renaturation of gelatin subunits during gelling depends on proline and hydroxyproline. The distribution of Gelatin's molecular weight and amino acid concentration affects all of its properties, including gel strength, viscosity, setting behaviour, and melting point. Gelatin containing a lot of amino acids tends to have a greater melting point and gel strength as a result [42]. The majority of gelatin purchased by the pharmaceutical sector is often used for the production of tablets, coatings for tablets, granulation, and hard and soft gelatin capsules (Softgels) where it improves the preparation's flavour and prevents oxidation. The research examining the use of gelatin-based biomaterials for the engineering of the heart, liver, skin, and cornea is covered in this topic. Gelatin and injectable fillers are discussed as possible treatments for wounds. Due to their aptitude for self-renewal and propensity to specialise into specific cell types, stem cells offer considerable promise for use in tissue regeneration. In order to increase the production of therapeutic proteins or to initiate gene silencing, respectively, gelatin-based delivery methods have been successful in delivering genes and siRNA. Overall, gelatin-based drug delivery methods have shown to be quite useful and adaptable in a variety of biological applications [43].

1.2.4 Chitosan

Arthropod exoskeletons are the primary source of chitin, although it may also be found in fungus, insects, and mushrooms. The exoskeleton of crab shells is made of chitosan, a glycosaminoglycan-like substance that is an N-deacetylated derivative of chitin and a member of the polysaccharide family. Due to its functionality, biocompatibility, biodegradability, and antibacterial qualities, chitosan has received extensive research as a natural biomaterial for several biomedical applications [44]. Biomaterials, or those made from biological sources, play a specific role in the search for the perfect dental materials because of their wide range of applications and innate biocompatibility. Chitosan is a relatively new substance that is mostly obtained from the exoskeletons of life, notably anthropods and fungus, etc. We are now able to use more biomaterials that are simple to adapt for human use, have fewer side effects, and have greater therapeutic results because of evolving technology and knowledge. Chitosan has more and more uses in medicine, thus research into its uses in dentistry also has to be pursued with greater zeal. Chitosan may be a genuine blessing in disguise for dentistry and find application in therapies ranging from preventative to regenerative dentistry in many of its specialisations, thanks to its features of being antibacterial, biocompatible, biodegradable, osteoconducting, etc. [45]. The most valuable by-product of chitin is chitosan, a natural polysaccharide. It is an effective microbiological agent with improved and increased qualities. After roughly 50% deacetylation, chitin undergoes 70% alkaline deacetylation to produce chitosan as shown in Fig. 4. The linear structure of chitosan is made up of a copolymer of glucosamine and N-acetylglucosamine. In the pharmacological, biomedical, cosmetics, hair care, acne treatment, and agricultural sectors, chitosan has a wide range of uses. In the pharmaceutical industry, it is used to accelerate the development of fruits and vegetables [46].

1.2.5 Hyaluronic Acid (HA)

Hyaluronic acid (HA) has been used more frequently recently as a biomaterial in both therapeutic and scientific purposes. Due to its great abundance in several tissues and ease of chemical manipulation, HA has become a prominent material platform for a

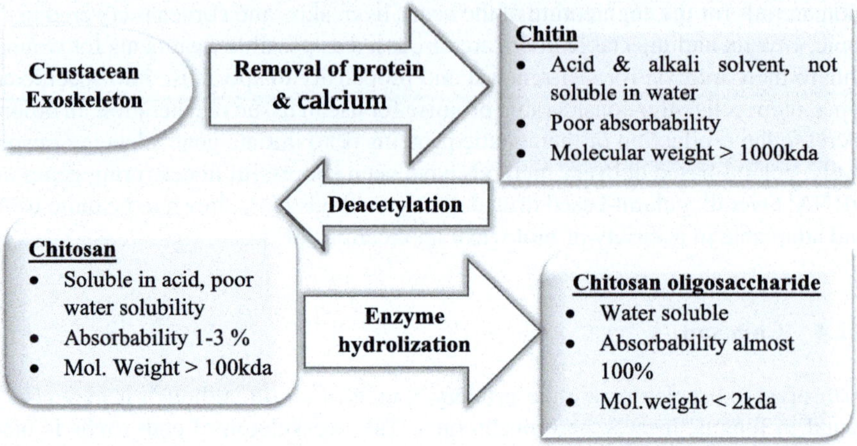

Fig. 4 Schematic diagram of formation of chitosan

variety of activities, including regenerative medicine, medication delivery, and scaffolding for cell culture. With its unique physicochemical characteristics and capacity to organise other matrix proteins, hyaluronic acid (HA) has long been recognised for its potential to influence tissue dynamics and remodelling. By activating cellular HA receptors, which may transmit signals that change cell survival, spreading, adhesion, and migration, HA can instead have a far more direct and focussed impact on how cells behave. Cells then alter HA by controlling its synthesis and breakdown using a specialised suite of enzymes. Taking into account all of these different mechanisms of control is necessary for the best design of HA-based biomaterials [47] (Fig. 5).

Nearly all physiological tissues and fluids contain hyaluronic acid (HA), a linear polysaccharide whose molecular weight (MW) varies depending on the type of tissue. Hyaluronic acid is sometimes referred to as hyaluronan. A sign of HA's biological significance and potential for therapeutic use is the fact that it is expressed basically everywhere. It is suitable for usage in a range of applications requiring conjugation or cross-linking because of its three orthogonal functional moieties (hydroxyl, carboxyl, and amide). These permit a variety of chemical changes [48]. Glucuronic acid and N-Acetyl-D-glucosamine are joined by alternating 1–3 and 1–4 glycosidic

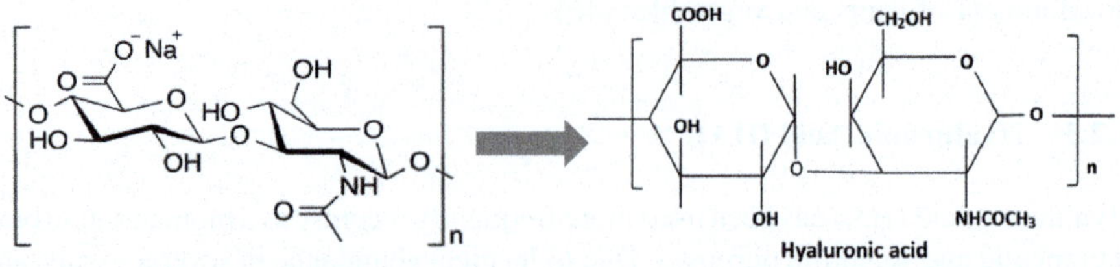

Fig. 5 Structure of hyaluronic acid

linkages to form the repeating disaccharide unit that makes up hyaluronic acid, a linear polysaccharide. The extracellular matrix of connective and epithelial tissues contains a glycosaminoglycan, which is its source. The material that grounds dermal cells and fibres into the skin is made up in part of hyaluronic acid, which is also present in the epidermis. In actuality, while being present throughout the body, hyaluronic acid is mostly produced by fibroblast cells in the skin, where it is primarily found [49]. Hyaluronic acid also has the ability to reduce inflammation and speed up the healing of wounds. The lower layers of the epidermis contain hyaluronic acid, which is thought to have a regulatory and modulatory function in the body due to its capacity to bind to proteins and cell surface receptors. The dermis's large hydrophilic polysaccharide, hyaluronic acid was once thought to contribute to the mechanical properties of skin by absorbing water. A variety of significant biological roles that this molecule plays in maintaining skin homeostasis are now being revealed by researchers [50]. The quality of the implanted cartilaginous extracellular matrix is improved by a bioink composition based on hyaluronic acid that permits 3D bioprinting. The ability to create 3D cell-hydrogel structures in 3D bioprinting that are strong enough requires the use of bioinks with high concentrations of polymeric components [51].

1.3 Tissue Engineering

Tissue engineering, a branch of biomaterial engineering, makes use of tissue synthesis to repair or replace whole tissues, such as skin, muscle, cartilage, bone, blood vessels, and many more. Tissue engineering, which is frequently used in conjunction with the term "regenerative medicine," is generally defined as the process of designing or directing the repair of tissues. However, the term can also refer to technologies used outside of the body, such as the construction of tissue constructs for in vitro research. The overlap results from the necessity of biomaterial technologies for the engineering of living tissues. Tissue engineering is an interdisciplinary field that aims to create biological tissues that improve, maintain, or repair tissue function or the function of a whole organ [52]. The following example illustrates how the basic concept of tissue engineering is represented in Fig. 6. A cell culture system or a bioreactor must be used to grow the isolated cells once they have been obtained from a source (allogenic, xenogenic, or autologous) (expansion in vitro). The enlarged cells must next be seeded into a carrier or matrix, which provides structural support and allows for the inclusion of the proper media (rich with nutrients and growth factors). The cells generate new tissues by differentiating, multiplying, and migrating to the carrier, which replaces the old tissues. The resulting tissue engineering construct is then used to create replacement tissue that is subsequently transplanted back into the patient [53].

Another way that nanotechnology is used in medicine is through tissue engineering (nanomedicine). By employing appropriate scaffolds comprised of nanomaterials and growth hormones, damaged tissues may be replicated or replaced. To build new tissues or organs to replace existing ones is the emphasis of the field of

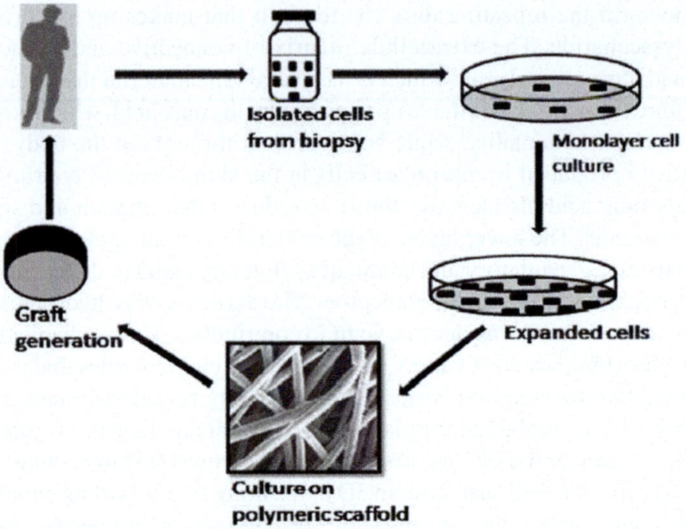

Fig. 6 Basic principle of tissue engineering

tissue engineering in nanotechnology. This initiative may take the place of currently used conventional therapeutic modalities like organ transplants or artificial implants. The host immune system can be suppressed, host tolerance can be induced, or immunomodulation of the tissue-engineered construct can assist avoid immunological responses in connection to these procedures [54]. Complex 3D working live tissues may now be printed using biocompatible materials, cells, and auxiliary components. With 3D Bioprinting, regenerative medicine is able to overcome the shortage of transplantable tissues and organs. When compared to non-biological printing, 3D bioprinting is more difficult due to considerations including material choice, cell types, growth and differentiation factors, and technological worries about the sensitivity of living cells and the formation of tissues. Combining engineering, biomaterials science, cell biology, physics, and medical technologies is important to find solutions to these issues. A range of tissues, including multilayered skin, bone, vascular grafts, tracheal splints, heart tissue, and cartilaginous structures, have been created and transplanted before using 3D bioprinting [55]. Tissue engineering is quickly growing into a sector with enormous potential, addressing particular medical requirements like organ failure or significant tissue damage. The objective of tissue engineering is to produce functional tissues that allow for the repair, upkeep, or enhancement of damaged organs or tissues. Diseases and disorders that might usually render a patient helpless or kill them can now be treated, thanks to tissue engineering. If spontaneous regeneration is impossible due to evolution, it allows tissue regeneration. Tissue engineering enables the body to repair itself, in essence. Tissue engineering is most frequently used to produce tissues that may be used to replace or repair bodily tissues that have lost some or all of their functionality. There are still many applications for tissue engineering, such as the creation of extracorporeal life support systems (such as bioartificial liver and kidneys), in vitro disease models, tissues for drug screening, smart diagnostics, and individualised medicine [56].

2 3D Printing

Numerous terms and descriptions that are similar to 3D printing are used to describe it, including additive manufacturing and fast prototyping. However, they all characterise additive manufacturing as the key concept that sets it apart from traditional subtractive techniques. The technology of 3D printing (3DP) allows for the creation of 3D solid things from digital files in any shape or geometry. The computer is given a computer-aided design (CAD) model of the design that will be produced. The term that broadly refers to printing on a range of materials, including polymers, plastics, ceramics, metals, and composites. In 1984, Chuck Hull invented the stereolithography technique, which involves curing photopolymers using UV lasers to build layers [57]. A 3D printer is made up of a variety of parts that work together to create the desired result from an input digital file. This is a list of a 3D printer's basic components: Extruder, hot end, print bed (tray), and filament. A computer file is converted into a real thing through the lengthy and intricate process of 3D printing. The procedure is broken down into the four steps listed below: Step-1: Modelling: CAD files may be produced either from scratch using a 3D modelling application or by starting with a 3D model produced by a 3D scanner. In any case, the application generates a file that is delivered to the 3D printer. Software separates the design into hundreds of horizontal layers, maybe thousands, as the process progresses. Create 3D models with 3D modelling software (such as CAD software), and download the model files (STL files). Step-2: Slicing: To slice the model, use a programme like ideaMaker. IdeaMaker creates a GCode file. Step-3: Printing: In order to prepare the printer for printing, upload the slice file and calibrate the printer. Step-4: Post-processing: Post-processing is the final stage in 3D printing. The post-processing for fused filament fabrication (FFF) 3D printing includes the following procedures, albeit not all of them must be carried out: removing the support, sanding, colouring, polishing, welding, and assembling everything. Some methods for forming the layers include melting or softening the substance, while others employ cutting-edge technology to cure liquid materials. Each approach has benefits and downsides of its own. Some frequently used technologies include the ones listed below: Selective laser sintering (SLS), multi-jet modelling (MJM), fused deposition modelling (FDM), stereolithography (SLA) [58]. For the manufacturing of polymer components, including prototypes and functional constructions with complex geometries, 3D printing methods are frequently employed. Due to their low cost, light weight, and processing flexibility, liquid or low melting point polymer materials for 3D printing are often employed in the sector. Being inert materials, polymer-based materials have traditionally played a significant role in biomaterials and medical device items. They helped the devices work effectively and provided mechanical support for numerous

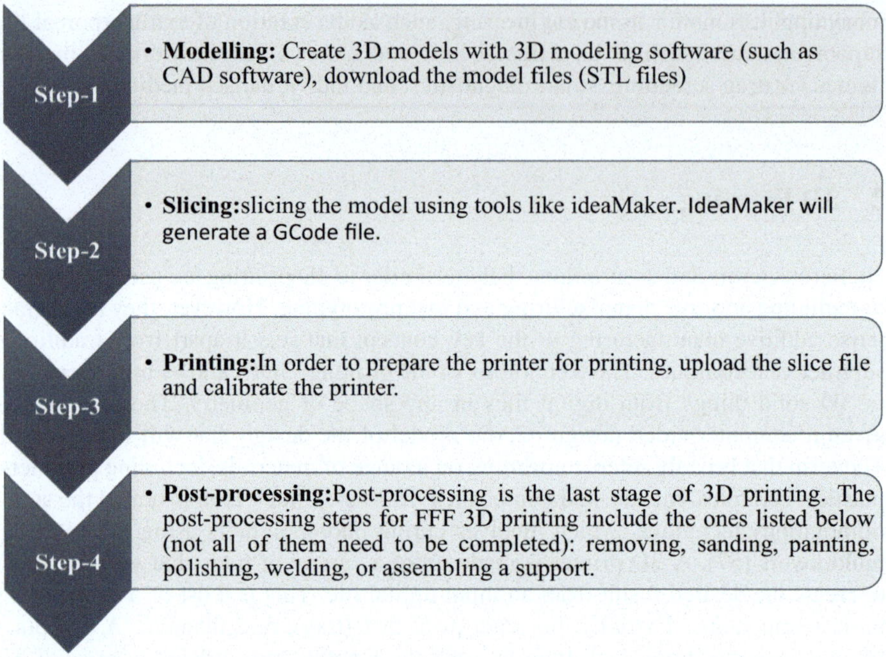

Fig. 7 Layout of a 3D printing model

orthopaedic implants [59, 60]. The natural structure of the skin can be more economically replicated by using 3D printing technology. On 3D printed skin, chemicals, pharmaceuticals, and cosmetics may be examined. Animal testing of products is thus unnecessary. The researcher will therefore be able to get accurate data by using a copy of the skin [61]. By using 3D printing technology, bony holes in the cartilage or bone brought on by illness or damage may be replaced. In contrast to employing auto-grafts and allografts, the goal of this treatment is to create bone, preserve bone, or improve bone function in vivo [62] (Fig. 7).

2.1 3D Bioprinting

3D Bioprinting (3DP) is one of the more advanced technologies which is out there where people are trying to fabricate scaffolds using this. The process of 3D bioprinting is slightly different from 3D printing. In 3D bioprinting, you will thus use cells together with the ink, whereas in 3D printing, you would simply print with standard ink and then seed cells on it. Extrusion-based, droplet-based, and laser-based 3D bioprinting techniques are grouped together as such. Bioinks that are filled with cells are patterned using laser radiation. The laser induced forward transfer, often known as the LIFT, is the method employed the most frequently in laser-based bioprinting. It is more common to use droplet-based bioprinting, which is also where 3D printing really began. This may be divided into two categories: drop-on-demand inkjet printing and continuous inkjet printing. Inkjet bioprinting offers the benefit of having incredibly high resolution and rapid printing. Moreover, it is relatively inexpensive. By using drop-on-demand procedures, it may introduce gradients in cell concentration. The most popular type of bioprinting is extrusion-based, and in this case, either mechanical or pneumatic pressure is utilised to extrude the Bioink from a nozzle. The mechanical force required to push the ink out might be a screw or a piston. Due to its capacity to control the hierarchical assembly of 3-dimensional biological structures for tissue building, 3D bioprinting is a method that has attracted significant interest for tissue engineering applications. In 3D bioprinting, functional components are positioned spatially and biological materials, biochemical, and live cells are carefully positioned layer by layer to create 3D structures. Three-dimensional bioprinting uses a variety of techniques, such as biomimicry, autonomous self-assembly, and microscopic tissue building blocks. Researchers are working on these approaches to build 3D functioning, living human structures that are suitable for therapeutic tissue and organ function restoration. The printing of fragile, living biological materials requires a major adjustment of printing technology designed for molten polymers and metals. Since extracellular matrix (ECM) components and many cell types have unique micro-architectures, it is difficult to mimic biological functions [55] (Fig. 8).

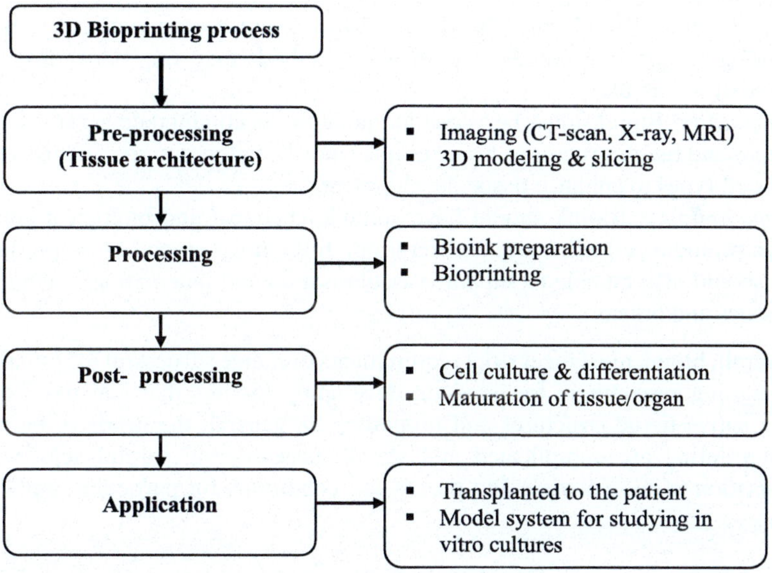

Fig. 8 Process of 3D bioprinting

2.2 Bioink Requirements for 3D Bioprinting

Bioink is a crucial component in the 3D bioprinting process, as it serves as a scaffold or support material for the living cells that will eventually develop into functional tissues and organs. The requirements of bioink for 3D bioprinting depend on several factors, including the type of tissue or organ being printed, the type of printer used, and the specific properties of the bioink itself. However, some general requirements of bioink for 3D bioprinting include:

1. *Biocompatibility*: Bioink must be biocompatible with the living cells being printed to prevent cytotoxicity or cell death. The bioink should also be non-immunogenic to prevent an immune response.
2. *Printability*: Bioink must be able to be printed with high precision and accuracy to create the desired 3D structure. The viscosity, surface tension, and other physical properties of the bioink can affect its printability.
3. *Degradability*: Bioink must be able to degrade over time to allow the cells to grow and mature into functional tissues. To maintain optimal tissue growth, the degradation rate should coincide with the rate of tissue synthesis.
4. *Mechanical properties*: Bioink has to have the right mechanical qualities to give the cells structural support as they develop and differentiate. The mechanical properties should mimic the native tissue being printed, such as stiffness or elasticity.
5. *Sterilisation*: Bioink must be sterilised to prevent contamination or infection during the printing process. Sterilisation methods may vary depending on the type of bioink used.
6. *Versatility*: Bioink must be versatile enough to accommodate a variety of cell types and tissue structures. Some bioinks may be tailored to specific tissue types or cell types to enhance tissue development.
7. *Functionality*: Bioink should have suitable chemical and biological cues that can promote cell adhesion, proliferation, differentiation, and ECM production. It should also be able to support vascularisation and innervation in engineered tissues and organs.

Overall, bioink must meet strict requirements to enable successful 3D bioprinting, and ongoing research is focussed on developing bioinks that can more closely mimic native tissue structures and properties. In general, the needs of bioink for 3D bioprinting are complicated and varied, necessitating careful selection and optimisation of biomaterials and processing conditions for particular applications [63].

2.3 Application of Natural Polymeric Biomaterials in Bioprinting

Natural polymeric biomaterials are widely used in bioprinting applications due to their biocompatibility, biodegradability, and ability to mimic the extracellular matrix of living tissues. Here are some of the applications of natural polymeric biomaterials in bioprinting:

1. *Tissue engineering*: Natural polymeric biomaterials are also used in bioprinting applications to regenerate damaged or diseased tissue. Natural polymeric biomaterials such as collagen, gelatin, alginate, and fibrin are commonly used to create scaffolds for tissue engineering applications. These scaffolds can be designed to match the properties of specific tissues, promoting the growth, proliferation, and differentiation of cells. For example, fibrin-based hydrogels have been used to create vascular networks, while alginate-based hydrogels have been used to regenerate liver tissue [64].
2. *Scaffold fabrication*: Natural polymeric biomaterials such as collagen, gelatin, alginate, and fibrin are commonly used to create scaffolds for tissue engineering. These scaffolds provide a 3D structure that can support cell growth, differentiation, and tissue regeneration. For example, collagen-based scaffolds have been used to fabricate cartilage tissue, while alginate-based scaffolds have been used to create pancreatic tissue [65].
3. *Drug delivery*: Natural polymeric biomaterials can be used as drug delivery vehicles in bioprinting applications. These biomaterials can be modified to encapsulate and release therapeutic molecules in a controlled manner, improving drug efficacy and reducing side effects. For instance, growth factors have been delivered to cartilage tissue using hydrogels made of gelatin [66].
4. *Vascularisation*: Natural polymeric biomaterials such as gelatin and fibrin can be used to create vascular networks within bioprinted tissues. These networks can improve nutrient and oxygen delivery to cells, leading to better tissue viability and function [67].
5. *Organ printing*: Natural polymeric biomaterials can be used to create functional tissues and organs through bioprinting. These biomaterials can provide an environment that is conducive to cell growth and differentiation, resulting in the development of intricate 3D structures [55].
6. *Organ-on-a-chip*: Organ-on-a-chip devices are also being developed using natural polymeric biomaterials. These systems can be utilised for disease modelling and medication screening because they replicate the structure and operation of live organisms. For example, collagen-based hydrogels have been used to create a liver-on-a-chip system [66].
7. *Wound healing*: Natural polymeric biomaterials such as chitosan and alginate have been used in bioprinting for wound healing applications. These biomaterials can be used to create scaffolds that promote cell migration, proliferation, and differentiation, leading to improved wound healing outcomes [68].

2.4 *Future Scope and Summary*

Natural polymeric biomaterials are materials that are derived from natural sources and are used in medical applications. These materials have gained considerable interest in recent years due to their biocompatibility, biodegradability, and low toxicity. They are employed in a number of medical procedures, including implanted medical devices, medication delivery, tissue engineering, and wound healing. One of the major advantages of natural polymeric biomaterials is that they can be easily modified to meet specific requirements. For example, the mechanical properties of these materials can be tuned by adjusting their composition or processing conditions. Natural polymeric biomaterials have demonstrated tremendous promise as building blocks for bioprinting because of their biocompatibility, biodegradability, and capacity to replicate the extracellular matrix (ECM) of real tissues. Aiming to improve patient outcomes in regenerative medicine, bioprinting is a fast-developing technology that shows promise for the creation of functioning tissues and organs. Collagen, gelatin, alginate, chitosan, and hyaluronic acid are some of the most popular natural polymers for bioprinting. These materials may be coupled with other materials, including synthetic polymers, cells, and growth factors, to improve their qualities. They have distinctive features that make them useful for a variety of applications. One of the key challenges in bioprinting with natural polymeric biomaterials is achieving adequate mechanical strength and stability in the printed constructs. To enhance the mechanical qualities of the printed constructions, researchers are creating novel cross-linking and curing techniques. The creation of novel bioinks that can be utilised to produce complex structures using various cell types and biomaterials is another area of ongoing study. Advances in bioprinting technology, such as the use of multi-material printing and advanced imaging techniques, are also expanding the possibilities for creating functional tissues and organs. Overall, the future of natural polymeric biomaterials for bioprinting looks promising, with continued advances in material science, bioprinting technology, and tissue engineering expected to lead to the development of new and innovative biomaterials and its applications.

References

1. Liaw CY, Guvendiren M (2017) Current and emerging applications of 3D printing in medicine. Biofabrication 9(2):024102
2. Kadakia RJ, Wixted CM, Allen NB, Hanselman AE, Adams SB (2020) Clinical applications of custom 3D printed implants in complex lower extremity reconstruction. 3D Print Med 6(1):29
3. Desu PK, Maddiboyina B, Vanitha K, Rao GK, Anusha R, Jhawat V (2021) 3D printing technology in pharmaceutical dosage forms: advantages and challenges. Curr Drug Targets
4. Gopinathan J, Noh I (2018) Recent trends in bioinks for 3D printing. Biomater Res 22:1–15
5. Ligon SC, Liska R, Stampfl J, Gurr M, Mülhaupt R (2017) Polymers for 3D printing and customized additive manufacturing. Chem Rev 117:10212–10290
6. Ramanath HS, Chua CK, Leong KF, Shah KD (2008) Melt flow behaviour of poly-ε-caprolactone in fused deposition modelling. J Mater Sci Mater Med 19:2541–2550

7. Skardal A (2018) Perspective: "universal" bioink technology for advancing extrusion bioprinting-based biomanufacturing. Bioprinting 18
8. Mobaraki M, Ghaffari M, Yazdanpanah A, Luo Y, Mills DK (2020) Bioinks and bioprinting: a focused review. Bioprinting 18
9. Guillotin B, Souquet A, Catros S, Duocastella M, Pippenger B, Bellance S, Bareille R, Rémy M, Bordenave L, Amédée J, Guillemot F (2010) Laser assisted bioprinting of engineered tissue with high cell density and microscale organization. Biomaterials 31(28):7250–7256
10. Guvendiren M, Molde J, Soares RM, Kohn J (2016) Designing biomaterials for 3D printing. ACS Biomater Sci Eng 2(10):1679–1693
11. Bedell ML, Navara AM, Du Y, Zhang S, Mikos AG (2020) Polymeric systems for bioprinting. Chem Rev 120(19):10744–10792
12. Gobbi SJ, Reinke G, Gobbi VJ, Rocha Y, Sousa TP, Coutinho MM (2020) Biomaterial: concepts and basics properties. Eur Int J Sci Technol 9(2):23–42
13. Isa ZM, Hobkirk IA (2000) Dental implants: biomaterial, biomechanical and biological considerations. Ann Dent Univ Malaya 7(27):35
14. Anselme K (2000) Osteoblast adhesion on biomaterials, review. Biomaterials 21(667):81
15. Anderson JM (2001) Biological responses to materials. Annu Rev Mater Res 31(81):110
16. Kohane DS, Tse JY, Yeo Y, Padera R, Shubina M, Langer R (2006) Biodegradable polymeric microspheres and nanospheres for drug delivery in the peritoneum. J Biomed Mater Res A 77:351–361
17. Troy E, Tilbury MA, Power AM, Wall JG (2021) Nature-based biomaterials and their application in biomedicine. Polymers 13(19):3321
18. Whitaker R, Hernaez-Estrada B, Hernandez RM, Santos-Vizcaino E, Spiller KL (2021) Immunomodulatory biomaterials for tissue repair. Chem Rev (in press)
19. Sofia S, McCarthy MB, Gronowicz G, Kaplan DL (2001) Functionalized silk-based biomaterials for bone formation. J Biomed Mater Res 54:139–148
20. 7 Different types of biomaterials and their applications that you should know. LinkedIn
21. Rockwood DN, Preda RC, Yücel T, Wang X, Lovett ML, Kaplan DL (2011) Materials fabrication from Bombyx mori silk fibroin. Nat Protoc 6(10):1612–1631
22. Murphy AR, Kaplan DL (2009) Biomedical applications of chemically modified silk fibroin. J Mater Chem 19:6443–6450
23. Sofia S, McCarthy MB, Gronowicz G, Kaplan DL (2001) Functionalized silk based biomaterials for bone formation. J Biomed Mater Res 54:139–148
24. Vepari C, Kaplan DL (2007) Silk as a biomaterial. Prog Polym Sci 32:991–1007
25. Omenetto FG, Kaplan DL (2010) New opportunities for an ancient material. Science 329:528–531
26. Altman GH et al (2003) Silk-based biomaterials. Biomaterials 24:401–416
27. Gu L, Shan T, Ma Y-X, Tay FR, Niu L Novel biomedical applications of crosslinked collagen. Trends Biotechnol
28. Parenteau-Bareil R, Gauvin R, Berthod F (2010) Collagen-based biomaterials for tissue engineering applications. Materials 3:1863–1887
29. Chattopadhyay S, Raines RT (2014) Collagen-based biomaterials for wound healing. Biopolymers 101:821–833
30. Wu X, Black L, Santacana-Laffitte G, Patrick CW (2007) Preparation and assessment of glutaraldehyde crosslinked collagen–chitosan hydrogels for adipose tissue engineering. J Biomed Mater Res Part A 81:59–65
31. Matthews JA, Wnek GE, Simpson DG, Bowlin GL (2002) Electrospinning of collagen nanofibers. Biomacromol 3:232–238
32. Matthews JA, Boland ED, Wnek GE, Simpson DG, Bowlin GL (2003) Compatible polymers electrospinning of collagen. J Bioact Compat Polym 18:125–134
33. Ahmed EM (2015) Hydrogel: preparation, characterization, and applications: a review. J Adv Res 6:105–121
34. Dinescu S, Albu Kaya M, Chitoiu L, Ignat S, Kaya DA, Costache M (2019) Collagen-based hydrogels and their applications for tissue engineering and regenerative medicine. In: Mondal M (ed) Cellulose-based superabsorbent hydrogels. Springer, New York, NY, USA, pp 1–21

35. Phull A-R, Eo S-H, Abbas Q, Ahmed M, Kim SJ (2016) Applications of chondrocyte-based cartilage engineering: an overview. BioMed Res Int 2016:1–17
36. Mariod AA, Fadul H (2013) Gelatin, source, extraction and industrial applications. Acta Sci Pol Technol Aliment 12(2):135–147
37. Bailey AJ, Paul RG (1998) Collagen—a not so simple protein. J Soc Leather Technol Chem 82(3):104–110
38. Ladislaus M, Kasankala YX, Weilong Y, Sun D, Hong Q (2007) Optimization of gelatin extraction from grass carp (Catenopharyngodon idella) fish skin by response surface methodology. Biores Technol 98:3338–3343
39. GMIA (2012). Gelatin handbook. Gelatin Manufact. Inst. Am. New York. http://www.gelatin-gmia.com/html/qanda.html. Accessed 22 Sept 2012
40. Mariod AA, Abdel-Wahab SI, Ibrahim MY, Mohan S, Abd EM, Ain NM (2011) b. Preparation and characterisation of gelatin from two Sudanese edible insects. J Food Sci Eng 1:45–55
41. Montero P, Gómez-Guillén MC (2000) Extracting conditions for megrim (Lepidorhombus boscii) skin collagen affect functional properties of the resultant gelatine. J Food Sci 65:434–438
42. Johnston-Banks FA (1990) Gelatin. In: Harris P (ed) Food gels. Elsevier Appl. Sci., London, pp 233–289
43. Nikkhah M, Akbari M, Paul A, Memic A, Dolatshahi-Pirouz A, Khademhosseini A (2016) Gelatin-based biomaterials for tissue engineering and stem cell bioengineering. In: Biomaterials from nature for advanced devices and therapies, pp 37–62
44. Jiang T, James R, Kumbar SG, Laurencin CT (2014) Chitosan as a biomaterial: structure, properties, and applications in tissue engineering and drug delivery. In: Natural and synthetic biomedical polymers. Elsevier, pp 91–113
45. Gayen K, Pabale S, Shirolkar S, Sarkar S, Roychowdhury S (2021) Chitosan biomaterials: natural resources for dentistry. Int J Oral Health Sci 11(2):75
46. Martinez-Ruvalcaba A, Chornet E, Rodrigue D (2007) Viscoelastic properties of dispersed chitosan/xanthan hydrogels. Carbohydr Polym 67:586–595
47. Wolf KJ, Kumar S (2019) Hyaluronic acid: incorporating the bio into the material. ACS Biomater Sci Eng 5(8):3753–3765
48. Cowman MK, Lee HG, Schwertfeger KL, McCarthy JB, Turley EA (2015) The content and size of hyaluronan in biological fluids and tissues. Front Immunol 6:261
49. Kavasi RM, Berdiaki A, Spyridaki I, Corsini E, Tsatsakis A, Tzanakakis G, Nikitovic D (2017) HA metabolism in skin homeostasis and inflammatory disease. Food Chem Toxicol 101:128–138
50. Maytin EG (2016) Hyaluronan: more than just a wrinkle filler. Glycobiology 26:553–559
51. Hauptstein J, Böck T, Bartolf-Kopp M, Forster L, Stahlhut P, Nadernezhad A, Blahetek G, Zernecke-Madsen A, Detsch R, Jüngst T, Groll J (2020) Hyaluronic acid-based bioink composition enabling 3D bioprinting and improving quality of deposited cartilaginous extracellular matrix. Adv Healthcare Mater 9(15):2000737
52. Bettinger CJ (2009) Pure Appl Chem 81(12):2183–2201
53. Vasita R, Katti DS (2006) Nanofibers and their applications in tissue engineering. Int J Nanomed 1(1):15
54. Fisher JP, Mikos AG, Bronzino JD (2007) Tissue engineering. CRC Press
55. Murphy SV, Atala A (2014) 3D bioprinting of tissues and organs. Nat Biotechnol 32(8):773–785
56. Mhanna R, Hasan A (2017) Introduction to tissue engineering. In: Tissue engineering for artificial organs: regenerative medicine, smart diagnostics and personalized medicine, vol 1, pp 1–34
57. Lipson H, Kurman M (2013) Fabricated: the new world of 3D printing, 1st edn. Wiley, Inc., Indiana
58. Canessa E, Fonda C, Zennaro M (2013) Low-cost 3D printing for science, education & sustainable development, 1st edn. ICTP—The Abdus Salam International Centre for Theoretical Physics
59. Caminero MA, Chacon JM, Garcia-Moreno I, Rodriguez GP (2018) Impact damage resistance of 3D printed continues fibre reinforced thermoplastic composites using fused deposition modelling. Compos Part B: Eng 148:93–103

60. Xin W, Man J, Zuowan Z, Jihua G, David H (2017) 3D printing of polymer matrix composites: a review and prospective. Compos B 110:442–458
61. Qian Y, Hanhua D, Jin S, Jianhua H, Bo S, Qingsong W, Yusheng S (2018) A review of 3D printing technology for medical applications. Engineering 4(5):729–742
62. Bogue R (2013) 3D printing: the dawn of a new era in manufacturing? Assem Autom 33(4):307–311
63. Gao Q, He Y, Fu J (2018) Chapter 4—bioink for 3D bioprinting. In: Wu X, Cui J, Hong K (eds) 3D bioprinting. Academic Press, pp 83–107
64. Gauvin R, Chen YC, Lee JW, Soman P, Zorlutuna P, Nichol JW, ... Khademhosseini A (2012) Microfabrication of complex porous tissue engineering scaffolds using 3D projection stereolithography. Biomaterials 33(15):3824–3834
65. Zhou X, Castro NJ, Zhu W, Cui H, Aliabouzar M, Sarkar K, ... Zhang LG (2018)
66. Lee JH, Park HJ, Kim K (2021) Bioprinting: principles and applications. Elsevier
67. Kolesky DB, Homan KA, Skylar-Scott MA, Lewis JA (2014) Three-dimensional bioprinting of thick vascularized tissues. Proc Natl Acad Sci 111(19):6940–6945
68. Gopinathan J, Noh I (2019) Recent trends in bioinks for 3D printing. Biomater Res 23(1):1–11

The manufacturer's authorised representative in the EU is Springer Nature Customer Service Centre GmbH, Europaplatz 3, 69115 Heidelberg, Germany. If you have any concerns regarding our products, please contact ProductSafety@springernature.com

Printed and bound by CPI Group (UK) Ltd, Croydon, CR0 4YY
01/10/2026
02225877-0015